D0887731

Remote Sensing:
A Better View

A New Collection By Duxbury Press.

The Man—Environment System in the Late Twentieth Century

General Editor
WILLIAM L. THOMAS
California State University, Hayward

REMOTE SENSING:
A Better View

ROBERT D. RUDD
University of Denver

DUXBURY PRESS
North Scituate, Massachusetts
A Division of
Wadsworth Publishing Company, Inc.
Belmont, California

Duxbury Press
A Division of Wadsworth Publishing Company, Inc.

ISBN-0-87872-068-5

L.C. Cat. Card. No. 74-75714
Printed in the United States of America
1 2 3 4 5 6 7 8 9 10 — 78 77 76 75 74

To my son Steve

Contents

List of Figures

List of Color Plates

Editor's Foreword

This valuable volume deserves a readership numbered in the millions. Only too rarely does an active practitioner in a rapidly expanding field of study address himself to persons who have only recently become aware of the topic. *Remote Sensing*, as a brief volume for the lay reader, is written interestingly in nontechnical language. It briefly explains the important developments in remote sensing and, in discussing some of their applications to problems of resource management, emphasizes their significance. Here is a volume whose purpose is to inform anyone why he should understand what remote sensing is and can do. The twentieth century citizen of the world well needs such understanding.

Dr. Robert D. Rudd, the author, is currently Professor and Chairman of the Department of Geography at the University of Denver in Colorado. He holds degrees from Indiana State University, the University of Wisconsin, and the Ph.D. from Northwestern University in 1953. Prior to his present position, he served on the faculties at Ohio University, University of Utah, and Oregon State University. His research and publications have focused on meteorology, climatology, physiography, and resources management. From 1968 to 1972 he was a member of the Commission on Remote Sensing of the Association of American Geographers, and in 1971-72 was chairman of that commission.

No volume such as this one has ever been published. Two compendia of technical articles or chapters recently have been compiled: Dr. Robert Holz of the University of Texas, Austin, edited a book of collected readings based upon 44 previously published articles, entitled *The Surveillant Science: Remote Sensing of the Environment* (Houghton Mifflin, 1973) and Drs. J.E. Estes and L.W. Senger of the University of California, Santa Barbara, organized and edited a volume, composed of chapters contributed by different authors, entitled *Remote Sensing: Techniques for Environmental Analysis* (Hamilton Publishing Company, 1973). Both of these larger works, however, are directed to those studying the techniques involved and assume a degree of interest and sophistication that most persons do not yet possess. The present volume is more philosophical than technical. It could well serve as a general introduction to a college course on remote sensing or as one of several volumes for a course on map and airphoto reading and interpretation offered for students in a wide variety of disciplines in the earth, biological, and social sciences interested in resources management, urban and regional development, and land use planning.

Those of the general public who are intellectually inclined and curious about the impact of new technological developments, government employees who will need to become familiar with these techniques, and professionals in a host of subject matter disciplines who will use the information gathered by remote sensing are just some of the intended audience for this volume. *Remote Sensing* is one of a series of volumes in the series, "The Man-Environment System in the Late 20th Century." The series as a whole is designed to present to the literate public those aspects of research in modern geography that should beneficially be known to a wider audience. Contemporary geography has much to contribute to man's grasp of the total man-environment system on the earth's surface. Each volume in the series consists of an integrated set of six to ten essays on a topic of social relevance currently at the frontiers of research. Their collective purpose is to widen your intellectual horizon through familiarity with a geographic viewpoint.

William L. Thomas

Preface

This is a book about a revolution. It is a relatively quiet revolution, although there is little about it that is secretive. Yet most people seem not to realize that it is going on, even though it has been given considerable coverage in the news media. Maybe the trouble is that the coverage has been in pieces and the totality of the revolution is difficult to visualize from its pieces. Or perhaps the terminology is the problem; reference to "remote sensing" in social conversation continues to elicit blank stares, despite the fact that the concept is not that new anymore. If the topic were extremely esoteric or one little likely to affect the average man, this lack of awareness would be easier to understand and to accept. But if remote sensing lives up to its potential, there will be few people in this world whose lives will not be affected by it. Moreover, the broad outlines of the concept are understandable to anyone who can read a newspaper.

The purpose of this brief volume is to introduce more people to remote sensing and to make them think about its potential, implicit in the pioneering efforts in research and application that have been carried out to date. That this is a book for the novice will be evident to the informed reader from my efforts to simplify and to emphasize the salient aspects of topics rather than their details. I have tried to convey a broad grasp of the total concept, believing that much of the mystery or confusion can best be dispelled by an organized look at the whole, even though that look be restricted in depth. There now are virtually entire libraries of papers, reports, and books which offer enough detail for the most sophisticated reader. Indeed, so prolific has technical publication been in this mushrooming field that the interested lay reader is at a loss to know where to start. Hopefully this volume and its deliberately restricted bibliography will provide a starting point. Finally, on a more personal level, I would simply like to communicate to others some of the fascination which the unfolding of this development, bright with promise, has held for me.

The Introduction and Chapter 1 of this volume are unabashedly intended to intrigue, in part to sustain the reader through Chapter 2. That chapter presents the basic whys and hows of remote sensing, so necessary to understanding applications and appreciating implications. From time to time a return to Chapter 2 from other chapters may be necessary to reestablish the reader's bearings. Although Chapter 1 addresses additional topics, these first three units about remote sensing seek to answer the question "What is it?" Chapters 3 and 4, and to some extent also Chapter 5, are concerned more specifically with "What can it do?" Lastly, Chapter 6 looks broadly at the question "Why should I be interested in it?" Here some of the negative aspects of remote sensing are raised and a modest attempt is made to compare both its assets and liabilities. In the hope that this volume will interest the reader in knowing more about the subject, a selected bibliography is included. The entries range from popular to technical and from very current to some that are rather "old" by remote sensing literature standards. Some of the latter are included because they do not assume as much background as more current efforts, and thus seem appropriate to this volume. An appendix also is offered which seeks to ameliorate what has been to date one of the more vexing problems for the initiate to remote sensing: where to get imagery. It suggests several possibilities for direct contact with sources.

I am indebted to many people for the experiences which led to this writing, but perhaps most to those with whom I worked on the A.A.G. Remote Sensing Commission. Among these, Professor Robert Peplies deserves special mention. Thanks are due many readers of parts of the manuscript, and especially Dr. John Estes, Dr. Donald Eschman, Dr. Paul Bock, Dr. Robert Amme, and Mr. Dan Drago for their comments and advice on what to change in the total manuscript. I made the final decisions, however, and they should not be held accountable for my mistakes.

Robert D. Rudd

Denver, Colorado
September 27, 1973

Figure I-1: A thermal image of an airport. Warmer features are light-toned, cool ones are dark. Note the numerous "shadows" of departed aircraft; the cement apron is cooler there. (Courtesy of Texas Instruments, Inc.)

Introduction

At first glance figure I-1 seems to be simply an ordinary aerial photograph of rather modest quality. It is a bit dark, as though underexposed, and lacks the crispness of detail that we have come to expect. A closer look reveals that many linear features such as runways are slightly misshapen, and a faintly perceptible series of closely spaced parallel lines overlies the whole picture. The more perceptive viewer, however, while puzzling over the unnatural brightness associated with the jet engines of the aircraft in the upper right corner, may realize what is really bothering him—there are aircraft shadows on many of the loading ramps without aircraft present to cast them.

Figure I-1 is not an ordinary photo made by a camera collecting light reflected from the scene. Rather, it is a picture made by collecting thermal infrared radiation emitted from the scene with a device analogous to, but distinctly different from, a camera. Various surfaces radiate differing amounts of thermal energy which can be collected and displayed as tonal differences on film, just as different amounts of reflected light give shades of gray. The cooler areas of the ramp which were in the shadows of now-departed aircraft radiate differently than the continuously exposed portion, producing the mysterious shadows which bothered you. The picture is an example of the products of a technique, in part new or greatly sophisticated, with exciting prospects—*remote sensing*. Some phases of it allow us to "see" things that our eyes, themselves, cannot see.

Extending the range of human vision is not new. A telescope allows us to look farther into space than we can otherwise see, and the microscope reveals miniature worlds too small for the unaided eye to resolve. But both of these devices merely help the eye to do its usual job, to use light for sensing objects. Some of the devices employed in remote sensing do not use light, however; they use other kinds of energy, as in the figure above. The images which result sometimes seem strange, but they represent an aspect of the object or scene that is no less real than the familiar one. They seem strange only because they are unfamiliar and we do not yet fully understand them.

Acceptance and utilization of the newer remote sensing techniques came slowly at first, and then mushroomed. In the mid-1960s studies employing the technique were only just beginning to appear in some of the professional journals. A few news media releases described aspects of the concept in piecemeal fashion, but they may or may not have used the term "remote sensing". Even among those who participated in the development of the technology or had been using remote sensing in a limited way, only a few perceived the full implications of the new techniques and visualized their possible potential. Toward the end of the 1960s, articles on remote sensing began to appear in popular periodicals, using the term itself and seeking to present an integrated view of the concept as well as some indication of its use. Today, articles and new releases abound. Even the most restricted professional journals include a few related articles, and several books and new technical journals devoted exclusively to remote sensing have appeared. Organizations such as the National Aeronautics and Space Administration (NASA) have virtual libraries of research reports on remote sensing. The concept is creating the potential for a revolution in many areas of endeavor where its application is beyond the preliminary stage. Nevertheless, to the average person the term is either totally unfamiliar or only vaguely identifiable.

Part of the reason for this unfamiliarity may be the events of the past few years, during which moon landings have become so prosaic that most people now are not certain how many times such landings have been achieved. The public has been seemingly buried in a plethora of space language, acronyms, and technological jargon until most have had a surfeit of terminology. And the term "remote sensing" does seem somewhat arcane. Today remote sensing has become so complex and sophisticated that few if any are master of all of its many facets. Yet an understanding of what remote sensing is, what it is doing, and some of its potential is within the reach of any interested person. The concept itself is not new at all; what is more, the basic principle is familiar to most.

Remote Sensing:

A Better View

Chapter   1

Toward a Greater
Breadth of Vision

Remote Sensing and Remote Sensors

Our senses, conventionally thought of as being five in number (although some of us are credited with having a "sixth sense") provide the mechanism by which we know our world. The parts of our bodies which allow us to experience sensations are one form of sensor. Some human sensors require physical contact with the subject. For example, you must touch velvet cloth to experience fully its softness. Your eye, however, senses shape, color, and size from a distance. In a manner of speaking, when you are merely looking at an object you are remotely sensing it; the amount of distance involved is unimportant as long as there is no physical contact between the object or scene and the sensor.

Of course, in usual parlance the term "remote sensor" refers to mechanical devices rather than human sense organs. An ordinary camera is probably the most familiar type of remote sensor. The analogy of the camera to the eye is commonly understood. Both employ light reflected from the object or scene through a lens and onto a light-sensitive surface to create an image.

Although most of us use cameras to record pleasurable events or sights we wish to remember, photographs also can provide much information for scientists familiar with their use. The information, both facts and figures, obtainable from photos makes them a very important data source. Although some remote sensors are capable of producing continuous current information at the same time as the sensor is operating (real-time data), most *record* the information in some form. The amount of usable information on a photograph is likely to be greater than on the continuously changing images of a display device which shows what the sensor is viewing. Remote sensors, then, are mechanical devices which collect information, usually in storable form, about objects or scenes while at some distance from them. Some, notably the camera, utilize visible light energy; others use different types of energy.

Somewhat less familiar than cameras are two other remote sensors—radar and X-ray machines. Both of these instruments obtain information from a distance, although with X-rays the distance may be little more than the thickness of a layer of skin and tissue. They both create beams of energy with which they sense target subjects or scenes, which may seem to differentiate them from cameras until you recall the use of flash attachments. The more important difference is the *nature* of the energy used in each system. For both X-rays and radar the differences in the wavelengths of the energy that they use give each system its unique advantage for certain tasks. Collectively, cameras, radar, and X-ray devices provide the capability for dealing with many kinds of information-gathering problems.

Radar came into operational use about 1940; X-rays were discovered near the turn of the century; the principle of the photographic camera is substantially older than both. If they are examples of remote sensing, the concept can hardly be called new. The concept in fact is *not* new, but the term "remote sensing" is. Two circumstances seemed to warrant the use of a new term. New devices for collecting information at some distance have been created, and those that we already possessed have become more sophisticated and consequently more useful. Collectively, these devices can provide a variety of "views" of a given subject and thus the opportunity to select the view best suited to a particular type of study. Additionally, where a short time ago photos were one thing and the image on a radarscope quite another, we can now produce radar images that look very much like photos. Moreover, we can produce photo-like images with still other devices that partially bridge the gap between conventional photos and radar images; and we also have expanded the limits of photography itself. Instead of two or three widely disparate sensing systems with no relatedness of application, we now have a quasi-continuous series of systems that can be used to complement one another's unique contributions. The use of complementary images, or ones better suited to a specific task than conventional photographic images, is providing new information and thus a more complete view than previously has been available. We are discovering features and phenomena that were there all along but unsuspected because we couldn't "see" them. We are collecting unfamiliar views of familiar features such as radar images of cities and fields (fig. 1-1) and finding that they help solve problems. The effort to visualize our surroundings more comprehensively employs different sensor systems in similar and related fashion and thus requires a descriptive term that will cover all: "remote sensing" is the term adopted.

Expanding Our View

Photographic Remote Sensing

The above description of what remote sensing is implies a great diversity of possible applications, but current applications focus mainly on man's cultural and physical environment. Until a decade or so ago, aerial photographs were the most familiar form of remote sensor data employed in most studies of the environment. The interpretation of conventional airphotos continues to be an important part of remote sensing. Moreover many of the techniques developed for the study of aerial photos are now employed in the study of images produced by sensor systems other than cameras.

The camera equipped with standard black and white or color film, however, records basically only that which the eye can see; and the need for additional views soon became apparent. By extending the sensitivity of film a little beyond the red end of the light spectrum, infrared film was obtained. Although much of the radiation from the sun is light energy, some of it is energy with wavelengths longer or shorter than those visible to the human eye. Infrared (IR) film is sensitive to some of the longer wavelength energy as well as to light energy. Because this longer wavelength energy reflects differently from foliage of vegetation that is live than from dead foliage, the film was useful during World War II in

Figure 1-1: Radar view of the Lawrence, Kansas, area. Field patterns and indications of crop differences are evident in tonal contrasts. The bright areas in and around the city are industrial and business districts. (Courtesy of David Simonett. NASA photo)

detecting instances where objects had been deliberately camouflaged with dead or artificial foliage. Color infrared film is especially good for this purpose because it depicts live healthy foliage in bright red (plate 1); thus, earlier versions of this film were called camouflage detection film. Once it was realized that a plant need not be dead in order to reflect infrared radiation less effectively, the film was put to use to detect distressed plants. Such application has proved to be of value to a variety of users. Another valuable asset of color infrared film is its haze penetration capability. The longer wavelengths of energy are scattered less by atmospheric contaminants than shorter ones; reflected red and infrared energy thus are recorded more effectively on the film than blue or purple light would be. Color IR film does not sense these shorter wavelengths; the ones it does sense penetrate the atmosphere more readily, thus affording a sharper image.

Although some success has attended efforts to use ultraviolet energy just beyond the other end of the visible spectrum, the results have not approached the successes achieved with infrared. The use of light energy plus the immediately adjacent

surface form are depicted on the radar image. (Courtesy of Goodyear
Aerospace Corp./Aeroservice Corp.)

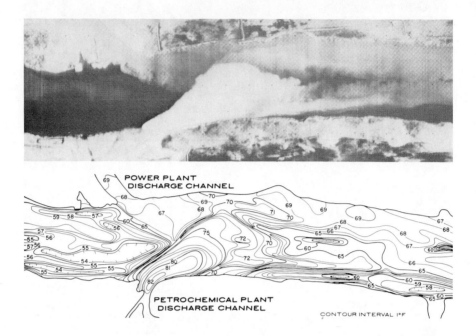

Figure 1-3: Infrared scanner detection of thermal pollution. The higher temperature of the effluent being discharged makes it immediately evident. Moreover, the actual extent of pollution is indicated by temperature identification. (Courtesy Texas Instruments, Inc.)

For example, thermal scanners have been used at night for determining the number of night-roaming animals. A better known use, however, is for detecting thermal pollution (fig. 1-3).

Both the radar and the thermal scanner use energy other than light, yet it is necessary that light be employed somewhere between us and the sensor if we humans are to see the information collected. An obvious answer is a scope, as used on radar-equipped ships and aircraft, but a device that provides only temporary "blips" has its shortcomings. A plan-like presentation of full information, such as a map or photo offers, is much more useful and is the format commonly sought. Developments of radar systems, such as Side-Looking Airborne Radar (SLAR), make it possible to create, scanline by scanline, an image of a scene which can be presented as a "photograph", hence the term "imaging radar". In analogous fashion, the data on emitted thermal infrared energy differences acquired by a scanner also can be presented in a photo-like format, enabling us to "see" these patterns. In neither of these cases, however, were the data collected by photography. The products of some other sensors are even less like photos and equally uninvolved with photography during the data collection. The question of what to call such photo-like renditions of scenes obtained by non-photographic sensors has been resolved by calling all of them *imagery.*

One other non-photographic sensor system should be mentioned here since it has something in common with both radars and thermal scanners, yet is different from each. The microwave radiometer senses radiation emitted from the scene as does a thermal infrared scanner but it senses much longer wavelength energy than the scanner—it is "tuned to a different channel." Microwave radiometers sense the same wavelengths as some radars. The signal received has several components though, some of which come from subsurface material at the scene. Thus, microwave radiometer data can provide information about soil moisture conditions.

A Broader View

An important point should have emerged in the past few pages, important enough to deserve reemphasis. We have been considering developments which provide us with a more complete view of our environment, a broader view. Broad, as used here, should not be confused with extent of area covered because this latter idea has to do with scale. Scale can become an important consideration as one assesses the utility of various remote sensing systems, but that is not the point we wish to raise here. A scale analogy, however, may be useful. Consider scale differences, for the moment, to be expressible in a vertical dimension. Generally speaking, the higher the altitude of the sensor the more area it can survey in a single view, and the smaller will be the scale of the resulting imagery. Conversely, with the same sensor nearer the ground, the smaller the area that can be surveyed at one time, and the larger will be the imagery scale. By varying scale, the views we obtain are different chiefly in the size of the features represented. On the other hand, suppose scale is held constant and we cause our several views to differ because of the sensors used. For example, consider variations in sensor type as being expressible in a horizontal dimension (fig. 1-4)—the different sensors arrayed along a horizontal line with those using longer wavelengths extending off to one side and those which use shorter wavelengths to the other side. As we change positions along *this* line (i.e., use different

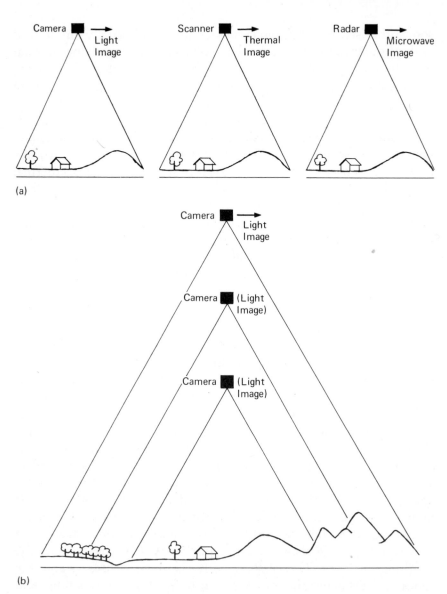

(a)

(b)

Figure 1-4: Change of view may be obtained by varying sensor type or by varying altitude. Changing sensor type (a) provides different kinds of imagery. Changing altitude (b) alters area covered, and therefore scale, but all images are the same kind.

sensors), we do not get the same kind of information merely at different scales; we get *different kinds* of information. The color infrared photo shows, say, variations in crop vigor in the wheat field, the thermal scanner image provides indications of subtle temperature differences across the field, a radar image may show irregularities in the surface more clearly, and the microwave radiometer indicates soil moisture characteristics. This information is not revealed as clearly or at all by conventional photography or by looking at the field while standing in it.

The topic we are discussing is man's newfound ability to obtain a more revealing view of things. We have broadened our vision beyond our corporeal ability to see. It is as though we now have several sets of eyes, each of which is sensitive to different energy wavelengths and thus capable of providing us with different information about a given scene.

A More Accurate View

There are many ways to use this newfound ability; the section following on applications provides some specific examples. More generally speaking, remote sensing has enabled us to discover features that were completely unknown; some are not yet fully understood. Other general uses, more prosaic but of considerable importance, include application in the search for *signatures*. The term "signature", as applied to imagery, refers to those visual characteristics of a subject which identify it and separate it from other similar subjects, much as differences in handwriting can identify different people. The identification of many subjects is simple, and one dominant characteristic precludes the need to investigate further. An airport runway pattern is a distinctive feature on an airphoto—its characteristic shape alone constitutes its signature, although some consideration is given to size. Other subjects may require a combination of criteria, and some pose a real challenge. It is important to some people to be able to distinguish between fields of wheat and fields of barley on remotely sensed imagery such as aerial photos, but this task proves more difficult than it may seem. Similarly, identifying tree species on conventional airphotos, especially when the trees are physically similar, is difficult; and mapping rock types in areas with little morphological or structural expression is another problem. There are cases in which an identifying signature is revealed simply by the use of the proper sensor: the use of an infrared scanner to detect thermal pollution, for example. But more ingenious is the concept of using a combination of imagery forms to identify a signature for a difficult subject.

The establishment of signatures is absolutely basic to many

aspects of remote sensing and much research effort continues to be devoted to it. Currently, signature research increasingly is oriented toward study of the characteristics of the received radiation itself rather than imagery produced from it. The broader view of remote sensing provides a greater variety of possible signature components and thus a greater chance for success with difficult problems. The simplified treatment presented here overlooks many problems, however, and may imply that the task is easy, which it is not. Similarly, the effort in the remainder of this chapter to convey an appreciation of the potential of remote sensing largely ignores the immature status of the technique and the unsolved problems which it faces. Both are as real as the potential.

Application

As we look into the question of what remote sensing can do for us, it is advisable first to establish some sense of perspective on one or two points to avoid misconceptions. There is the question of what might be called application status, or "To what extent is this actually being done?" A survey of the literature provides examples of at least three distinctly different categories of status. First, there is what is known as *operational* status, in which remote sensing is being and has been applied routinely on a daily or continuing basis. There is no question of whether or not remote sensing will do the job here because it is doing the job. Secondly, there is *experimental* status: "Under such and such conditions, the following results were obtained in the instance being reported here—and research is continuing." The question remains with this status whether the same results will be forthcoming from duplicate or similar experiments. Presuming a competent original effort, an experiment of this status is indicative of what is likely. Thirdly, there is a status which we can call *theoretical* here although the term may be questionable. It refers to an idea for application which may have been inspired by operational or experimental evidence but which has not yet been tested experimentally. Such an idea is a possibility but as yet is no more than the proverbial gleam in the progenitor's eye. Examples of each status are included in this discussion of application and the reader should distinguish among them.

Additionally, a point of clarification should be made regarding sensor *platforms*, another bit of remote sensing jargon considered necessary because sensors are transported by a variety of means, including aircraft, spacecraft, balloons, ships, buoys, and even automobiles. The discussion in this volume emphasizes only aircraft and spacecraft, the most familiar examples, but mixing just these two is a common occurrence. The use of some sensors in spacecraft has only reached the experimental stage, or may yet be theoretical, and it should not be assumed that all sensors which operate from aircraft also are being employed in spacecraft. Many are, and the effort to narrow the gap continues, but it is likely that some sensor systems will never become operational in spacecraft. Indeed, the point will be made or implied here more than once that the complementary use of data from airborne and spaceborne sensors is the desired goal, rather than the exclusive or dominant use of one or the other. At present it appears that for many tasks the scale or type of imagery needed would be obtainable from spaceborne sensors only through unreasonable effort or expenditure.

The Variety of Potential Application

What, then, are some examples of what remote sensing can do? Despite the relatively short time since remote sensing first became recognized as a viable research technique, attempts at applications are incredibly numerous and varied. The use of remote sensing by the National Weather Service has been going on now for so long that many do not associate the familiar phenomenon of the weather satellite with the term. Application of remote sensing techniques to the study of weather is a good example of operational use. Observation of the movement and development of storms on satellite imagery has been a routine procedure for a number of years, and this information has proved to be especially valuable in the forecasting of severe storms. The first U.S. weather satellite was launched in 1960, and such satellites have provided continuous views of cloud patterns ever since (fig. 1-5). Cloud patterns, however, are only one type of weather information obtainable from current satellites; a more detailed look at remote sensing and weather is offered in Chapter 3.

Agriculturalists are finding many applications for remote sensing. The early detection of plant disease or of inroads by insect pests through the use of several sensor systems would reduce losses by permitting remedial action to be initiated more promptly. To prevent plant diseases or curtail destruction of selected crops by

Figure 1-5: Cloud patterns which reveal the presence of a cyclonic storm as portrayed by an image dissector camera on the Nimbus 4 satellite. The cold front (1) is characterized by cumulus clouds (2) with some bright cells (3) of considerable vertical extent. Stratus clouds (4) can be seen along the warm front (5) and in the occlusion (6). The arrows show the abrupt end of cirrus bands which indicate the location of the jet stream north of the warm front. (NASA photo)

insect pests, the established practice is to spray the fields of crops routinely several times a season. In areas that have extensive acreages in such crops, unnecessary spraying might be eliminated by monitoring the fields by remote sensing for evidence of crop stress. Recently the U.S. Midwest experienced severe outbreaks of corn blight, and experiments are being conducted to determine whether the problem could be detected by satellite sensors. The launch in July 1972 of the first Earth Resources Technology Satellite (ERTS-1) hopefully presages a new era in crop disease detection. The satellite returns to a position over the same location on the earth every 18 days, thus providing frequent opportunities to watch for indications of plant stress. Although some details may not be available from ERTS imagery because of scale, aircraft overflights could be used to pinpoint trouble spots once a problem is indicated from space. Improved predictions of yields through monitoring of crop vigor is another hoped-for product from such satellites, perhaps from succeeding models if not from ERTS-1.

Microwave radiometers have shown experimental capability for indicating the water content in snowpacks, and can provide such indications about the total area scanned. Because snow cover depths are nonuniform, the sensor can provide better water supply forecasts than the estimates based upon on-the-spot sampling by surveyors at various points of the surface. Thermal infrared scanners are able to locate small, smoldering fires in forests and are used operationally in fighting forest fires; by penetrating the smoke they are able to locate hot spots and delineate the exact fire perimeter.

General procedures for inventorying natural resources are expected to be modified greatly through the addition of the newer remote sensing techniques. What was once a problem in airborne photography for mapping, such as cloud cover in inclement weather, is yielding to sensor systems not affected by clouds. A six-year effort to collect airphotos for mapping a huge area in Brazil produced acceptable coverage of only half the desired area, whereas recent overflights employing radar (SLAR) very quickly produced total coverage. A United States firm was recently engaged by Algeria to carry out an $8 million resource survey by airborne remote sensing to cover the entire country. Algeria, as a result, may have one of the most complete and up-to-date surveys of its resource base of any nation in the world.

Ecology and pollution control may benefit also from remote sensing applications. The use of thermal infrared and microwave radiometric sensors to obtain urban heat-island profiles provides a basis for subsequent modeling procedures which may be able to predict the effect a new freeway would have on urban climate. Some types of imagery have been found experimentally to depict aspects of neighborhood quality, while sequential coverage gives even more effective monitoring of the effects of urban sprawl. Oil spills are another problem of increasing concern. Often conventional information-gathering procedures fail to depict the extent of oil spill or keep track of it, but several remote sensing techniques have indicated potential for the task. Oil slick luminescence is detectable by ultraviolet sensors; thermal scanners and microwave radiometry also can detect oil spills. Some of the oil pollution at sea is not accidental, and the U.S. Coast Guard would like to employ airborne or spaceborne radiometers to monitor pollution ordinances. Some of these sensors are especially effective at night or in clouds and fog, the preferred conditions during which ship bilges are flushed.

Finally, at least some mention of the "black box" category of application should be made. Thermal infrared sensing of the human body has produced some surprises. Skin temperature patterns have revealed unexpectedly warm or cool spots believed to be associated with such diverse causes as subcutaneous tumors or circulatory irregularities, casting this kind of remote sensor in the role of a diagnostic aid. On a different tangent, news media have made references to law enforcement applications. Light amplification devices which are capable of enhancing the seemingly nonexistent light on a dark street several thousandfold, so that objects a half mile away are visible at night, are being tested in some urban areas. Reports of experiments with some extremely long wavelength radar-like devices have almost a science-fiction character. The penetration capability of long wavelength energy allows these devices to peer through concrete walls. To be sure, the fuzzy and uncertain "vision" leaves many doubts about what the signals being received actually show. Still, that such devices have reached the stage that they are being tested and *reported* raises disquieting questions about extant equipment that is subject to security classification.

Collectively, then, the devices of remote sensing have application to many and diverse tasks. In all instances, however, the same fundamental commodity is being sought—information.

The Potential for an Information Explosion

The Present Data Base

It is easy to be misled about how well informed man is regarding his world. History records the 15th and 16th centuries as the "Age of Discovery", and that was a long time ago. Surely, enough "last frontiers" have been identified; can there be any more on earth? A veritable army of scientific investigators has been mustered during the 20th century; is there anything left that has escaped their attention? Libraries are bulging with information about man's world; indeed, those libraries have atlases whose maps portray the religions, economies, minerals, crops, natural vegetation, and so on, of the entire world. To be sure, there are more names or symbols on the maps for some parts of the world than others, and the distribution patterns of whatever is being mapped in places like the Amazon basin or central Africa are much less complex than for the United States or Europe; but perhaps that is just the way things are. The faint suspicion that arises as you look at the atlas version of an area that you know well is allayed by the realization that at such small scales one must expect generalization. It is when more detailed information is sought that suspicions become more firmly rooted. A part of a detailed map is blank with the word "unsurveyed" across it; or the map information is so old that the utility of the map is limited; or even the most detailed map available may not provide the kind of information needed; or there simply is no detailed map of the area in question.

Maps are only one form of data summarization or storage, but the same or similar problems emerge with other forms if you probe deeply enough. No one really knows how many acres this year are planted to wheat in the United States. Any figure stated can only be an estimate. Not every wheat field is visited by inspectors nor is the entire country photographed from airplanes several times each year in order to revise the estimates of the harvest. This year's wheat crop will remain an estimate until the harvest is in and measured; and so similarly with corn, oats, and other crops. If the U.S. crops are at best estimates, the world crops can only be guesstimates.

Figures on the volume of timber in the forests of this country are continually being revised as more detailed studies of sample areas indicate inaccuracies of earlier assumptions, as forest fires and disease or insects extract their toll, and as growth increments and annual harvests are considered. In the case of forests there is an additional consideration—we have never really known the exact extent and nature of the resource since much of it consists of mixed stands. The fairly good idea of what we have in the United States remains but an estimate. The content of the world's forest resource is speculative, because vast areas of tropical forest are imperfectly known.

How is land use changing? What proportional changes are occurring in cropland, rangeland, and forests as urban sprawl and transportation facilities encroach upon them? Should we regulate these things? To what extent and in what manner? We have information on what is happening on the local scenes, but we need to know the overall national picture—and the international picture.

There are obvious difficulties to be overcome in any effort

involving international cooperation and information exchange, but they are surmountable. The World Meteorological Organization is the proof: international exchange of current weather information is routine. Weather may be local but its causes are global, and it has become increasingly clear that improved forecasting will be dependent upon improved data coverage. Numerical weather forecasting, in which the capacity and speed of the computer are put to use to obtain a more complete summary of conditions, requires some global data input, especially as a basis for extended forecasts. The ocean of air which surrounds us is vast indeed, covering millions of square miles of area and having a depth of tens of miles; the present observational network for monitoring atmospheric changes is inadequate for the task. Weather satellites have improved the situation substantially, but some of the needed information is being obtained from them only on an experimental basis. We know enough now to believe that the Arctic plays a far more important role in global weather than previously was imagined, but our knowledge of the Arctic is minimal. The stations which survey conditions over the oceans, which cover nearly three-fourths of the globe, are only a fraction of the number of continental surface observation stations.

The Need

There are many situations in which a better data base would improve our ability to assess our resources, make better use of them, understand our environment, and so on; but sooner or later a question must be faced—is it necessary? Two or three considerations suggest that it is. For one thing, the population explosion, although no longer news, is no less real. People in unprecedented numbers will be with us and they will need to be fed, clothed, and housed. Moreover, they will demand more than these three essentials. Signs of strain indicate that present methods, policies, and monitoring procedures already are inadequate locally and nationally. With the numbers that are in prospect, it will not be enough simply to do better what we have been doing.

Aside from numbers alone, the rising standard of living which all people want to experience will require more of everything *per person*. The requisite technology to satisfy such needs and numbers could make the ecological impact we have observed to date pale by comparison. And finally there is the accelerated rate of change with which things can happen now. Our capability for modifying the earth's surface has so increased that landscapes can be changed overnight. Even though forethought is given to the possible effect of change, the real test is the observed results which follow. When changes are so rapid, so extensive, and so numerous that their cumulative effect can produce unforeseen results, the kind of monitoring which has served in the past will no longer suffice.

So what we are going to need (if indeed we do not already) is quantities and types of information heretofore unimagined. If we are to have an adequate appraisal of our resources on which to base plans for more ordered use, if we are to be able to foresee incipient problems soon enough to make remedial action worthwhile, and if we wish to understand the natural environment well enough to be able to prepare for or modify its diverse moods, we must improve our ability to collect information. Remote sensing offers ways to help us.

The Contribution

The broader view of remote sensing will be needed, not only for what additional information it can provide about the known, but especially for its probable application to presently unknown but sure-to-be very sophisticated future problems. The more complete view should provide a sounder basis for making future decisions. And remote sensing from satellites offers an additional plus: an extent of coverage not approached by non-orbiting systems. First of all, the satellite's altitude enables the sensors to scan a large area at a glance. For example, an area 100 nautical miles by 100 nautical miles is now depicted in a single frame of imagery from the recent earth resources satellite. Although the small scale on such imagery may limit its use in the study of detail, there are tasks for which detail is merely a bothersome addition; for such studies it is more important that one space photo covers the same area as thousands of airphotos. In addition, orbiting satellites continue to send information, making repetitive or sequential coverage of an area available. The march of the seasons becomes visible in the appearance of snow cover, the changed appearance of areas with seasonal vegetation foliage, the formation and breakup of ice cover on water bodies, and the rhythm of field crops. The opportunity to watch a whole continent change throughout the year provides information of an extent never before available.

The amount of information that has been returned from weather satellites is astounding (plate 2); tens of thousands of images have been produced from the data collected by a single satellite in some cases, and quite a number of such satellites have been sent aloft. Without help we will not be able to make use of all of the information that future satellites will provide. Automated reading of taped data or scanning of imagery from some satellites has been proved possible, and computer map printouts of selected features have been produced. Automation is not likely to render the skilled interpreter obsolete, but it can reduce his workload by handling bulk tasks and pointing up features which need closer attention.

An opportunity for a "quantum jump" in data acquisition is afforded by remote sensing and orbiting sensor platforms. Continuous temporal coverage, a variety of scales through the appropriate use of aircraft and spacecraft, and the variety of views which the several sensors provide offer the opportunity to move toward a new era in man-environment interrelationships, at a time when one seems very necessary. Remote sensing is not a panacea, however. It simply is an additional but very powerful tool to be added to those which already are in use. And finally, it will not provide the solutions to problems; but remote sensing *can* provide the basic information necessary for arriving at solutions.

Chapter   2

The Basics of Remote Sensing

An understanding of the fundamental principles on which the concept of remote sensing is based and a general awareness of how some of the equipment works are not prerequisites to reading about what the technique can do. People read about applications daily in the news media without such background. But it is difficult to appreciate fully just how startling is this development unfolding before us without some foundation for understanding it. Similarly, the potential of the technique and the implications of such capabilities are more difficult to visualize when the whole is only perceived as a group of vaguely related parts. Hopefully this chapter will explain some of the parts and place them into an organizational framework to which the reader may return periodically from subsequent chapters.

Electromagnetic Radiation

The term "remote sensing" is most frequently used today to mean the collection of information from a distance through the use of radiant energy. This interpretation is at once both more broad and more restricted than some would have it. Inclusion of the study of subjects that are not at truly great distances makes the use of the word remote questionable for some; others feel that media other than radiant energy (sound, for example) should be included. The facts remain, however, that the term has experienced wide use in remote sensing studies which involve the full gamut of distances, and that some form of radiant energy has been used to collect information in most such cases.

Radiation is one of the three commonly recognized modes of transference of energy, the others being conduction and convection. Radiation is unique among the three in that radiated energy can be transferred across free space as well as through a medium such as air. Thus it is that sunlight, a form of radiant energy, crosses the gulf of emptiness between the sun and the earth. The energy itself originated in the complex activities that go on at and below the surface of the sun.

Physicists tell us that radiation is generated when an electrical charge is accelerated. For most of us this efficient explanation is inadequate, raising more questions than it answers. For amplification, suppose that the electrical charge in question is a part of an atom such as an electron. Various natural occurrences can cause electrons to change position within the structure of the atom. Certain types of position changes result in the requisite acceleration, and radiation is given off as a result. The radiant energy emitted is called *electromagnetic energy* in recognition of the fact that it has both electric and magnetic components.

In the example above, we assumed the accelerated electrical charge to be an electron. Larger particles such as molecules, if electrically charged, also will produce radiation when accelerated. It is still electromagnetic energy, but it has different characteristics. One of the differences is the *wavelength* of the energy, and we use this characteristic to distinguish one type of radiation from another. Another characteristic that can be used to differentiate electro-magnetic energy types is *frequency*. Although it is not technically correct to do so, the familiar phenomenon of ocean waves is commonly used to illustrate the meaning of the terms wavelength and frequency. The distance between adjoining wave crests is wavelength and the number of crests passing a given point during a given time is frequency. Since all electromagnetic energy—unlike ocean waves—travels at the same speed, longer wavelength energy will have fewer crests passing a point per unit of time than shorter wavelength energy. In other words, the greater the wavelength the lower the frequency.

The Electromagnetic Spectrum

The reason for the preceding several paragraphs is to provide some basis for the introduction of an organizing concept which will illustrate the interrelatedness of the several facets of remote sensing. The concept is the electromagnetic spectrum.

The two examples of sources of electromagnetic radiation discussed here, the electron and the charged molecule, may constitute a misleading sample of a very large population. Any known substance having a temperature above what is known as absolute zero ($-273°$C or $0°$K)* is considered capable of emitting radiation if circumstances are right, and the temperatures below absolute zero are not believed to exist. The point is that there are so many kinds of known substances and so many possible variations of conditions in the universe that there is an enormous range of radiation categories possible. They display wavelength differences which vary from hundreds of kilometers down to unimaginably short dimensions. In fact, we visualize a continuum of varieties of electromagnetic energy between these two extremes and the total array is called the electromagnetic spectrum (fig. 2-1). At one extreme is long wavelength radiation such as radio waves; these longer wavelengths are typified by lower frequency as mentioned before. At the other extreme the higher frequency forms such as X-rays and gamma rays have wavelengths so short that they are expressed in units unfamiliar to most of us. For example, X-ray wavelengths may be expressed in angstroms, which are units of length equal to one hundred-millionth of a centimeter.

As we look at a diagram of the electromagnetic spectrum, we can begin to put into order some of the things that were discussed earlier. Each of the sensors introduced in Chapter 1 utilizes the energy of a particular part of the spectrum and this fact can be helpful in distinguishing among the sensors and their capabilities. Indeed, the spectrum is commonly thought of as being divided into regions. Wavelength dimensions are used to bound the regions, although usage implies that most boundaries are not thought of as being precise. Boundaries are commonly visualized as zones of transition from one region to the next.

So it is, then, that we identify a region of the spectrum called the visible or light region; it includes electromagnetic energy with

---

*A temperature scale called the Kelvin scale is commonly used in science. To change a temperature from °C to °K you add 273. 0°C is thus 273°K.

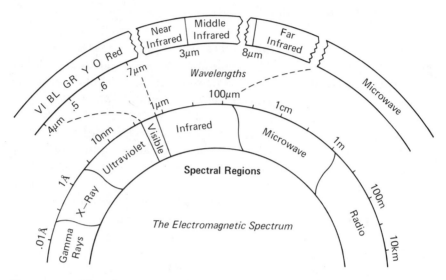

Figure 2-1: The electromagnetic spectrum.

wavelengths from a bit less than four-tenths of a micrometer (.4μm) to nearly .8μm. Although light has much longer wavelengths than X-rays, these wavelengths are still very short, since a micrometer equals one millionth of a meter. The position of the visible region of the spectrum is somewhat left of center on the diagram (see fig. 2-1) but the diagram deserves an additional look. Notice that the wavelength values are indicated at marks that are equally spaced but the values themselves change by a factor of 10 for each successive mark. It is a highly compressed diagram and none of the increments on it are intended to represent actual wavelengths. A not uncommon reaction of people realizing for the first time just what a spectrum diagram shows is surprise at the smallness of the visible (light) region. The miracle of human vision seems somehow diminished by this expression of its limitation.

The eye and the camera equipped with conventional film use the electromagnetic energy of the visible region of the spectrum. Adjoining this region on the longer wavelength side is the infrared region. Although we can use cameras with special film to sense in the immediately adjacent portion, the majority of the infrared region requires a different type of sensor, such as the thermal scanner. The infrared region constitutes a much larger portion of the spectrum than the visible, extending from wavelengths slightly less than 1μm to near a millimeter; the longer waves thus are approximately a thousand times the length of the shortest.

In this introduction to the spectrum, only one other region

will be mentioned, the microwave region. The wavelengths in this region are still longer than those of the infrared region, beginning at about a millimeter and extending to beyond a meter. These are the wavelengths used by radars and the microwave radiometer. Although they utilize energy of similar wavelengths, the two instruments are quite different; hence, the kinds of information they collect are different.

As we take a closer look in the sections which follow at the sensors which utilize the energy of the different spectral regions, readers who are bothered by the narrowness of the visible part of the spectrum may find reassurance. While man's visual capabilities may be limited, his ingenuity is not. He is learning to broaden his ability to see with the development of mechanical "eyes" that utilize a much broader portion of the spectrum.

Sensing With Light

There is a fundamental relationship between the temperature of a radiating body and the wavelengths of the radiant energy; as the temperature of the body increases, the dominant wavelengths of the radiation become shorter, and vice versa. A high-temperature surface thus radiates predominantly shorter wavelength energy; conversely, a low-temperature surface radiates mostly longer waves. Radiation from the sun, which has a surface temperature near 6000°K, shows a peak in the wavelengths of the visible region of the spectrum. Although some solar radiation is of shorter (ultraviolet) wavelengths, most of the remainder is infrared. Solar radiation is thus considered to be predominantly short wavelength energy, although there is no fixed division point between short and long wavelengths. Conventional photography, which is an important part of remote sensing, utilizes only visible light energy, either natural or artificial. The part of the spectrum we are concerned with for the moment is thus between the wavelengths of about .38μm and .78μm. It is interesting, although not surprising, that the human eye has its greatest spectral sensitivity in the middle of this region near the same point at which solar radiation is greatest.

Conventional Photography

The basic principle of the most familiar remote sensor, the conventional camera, is well known. Consideration of its capabilities and limitations with reference to the spectrum may provide some additional perspective, however. Standard black and white aerial film is sensitive to wavelengths of approximately .36μm to .72μm, essentially the same as those to which the eye is sensitive. Although this similarity is intentional, so that the image obtained will be a familiar one, difficulties are encountered in trying to extend this range very much for a conventional camera and film system. Optical glass presents a problem for wavelengths of less than .36μm, limiting the camera's utility in the ultraviolet. Of course we do go beyond .72μm with cameras, but we must use infrared sensitive film. Rather than expanding the number of wavelengths sensed, we are obtaining some interesting results by restricting the number. That the visible region of the total electromagnetic spectrum is a spectrum within itself is well known, and it is possible to identify individual bands within the region by wavelength. The light energy which creates the sensation of blue in the eye has wavelengths near .4μm, green about .55μm, and red about .7μm. With filters, it is possible to restrict the wavelengths which reach the film so that a photo can be made with only one band such as "green light" or "red light" (fig. 2-2). With black and white film, the photo is still black and white, but subtle differences between it and one made with the broader range of wavelengths are discernible. Comparing photos of the same subject taken at the same time but with different bands of light has indicated that some tasks are better served by one band than by others.

The use of multiple layers of emulsion on the film base, each sensitive to different wavelength of light and containing appropriate dyes, made possible the color film so familiar today. Working use of the film has proved its superiority to black and white for many tasks, notably because the image is more like the real world as we are accustomed to seeing it. It is especially valuable for tasks in which the distinction between like subjects is very subtle; it is estimated that the eye can distinguish 100 times as many color and tone combinations as gray scale values. It is not true, however, that all tasks are better served by color photography, and black and white photos will remain the workhorse for many remote sensing efforts.

Conventional photography is not likely to be replaced by non-photographic remote sensing techniques in the foreseeable future. A comparison of examples of the best of each reveals a basic reason. The resolution or landscape element resolving capability—the fidelity of detail reproduction—on conventional photography is typically superior to that obtainable by other sensor

Figure 2-2: View variations resulting from use of different spectral bands. The top image was made with "blue" light, the second with "green" light, the third with "red". The bottom image was not made with light; it is a thermal IR image. (Courtesy of the Bendix Corp., Ann Arbor.)

systems. Conventional photography also affords a more familiar view and is usually less expensive. Sophistication of camera systems and films, innovative use of film/filter combinations, and other technical advances should enable photography to remain an important facet of remote sensing.

The Near Visible

The terms "infrared" (IR) and "ultraviolet" (UV) imply the proximity of these regions of the spectrum to the visible region. The prefix "ultra" means beyond and the ultraviolet region is the one immediately beyond the violet portion of the visible region on the shorter wavelength side. "Infrared" literally means below the red, energy with frequencies lower or with wavelengths longer than the red part of the visible region. Although the eye can not detect either ultraviolet or infrared, both were identified in the laboratory nearly two centuries ago.

Photographic Infrared

The energy that we utilize in taking infrared photographs is reflected solar radiation, but only those wavelengths which are quite close to visible light. Although some infrared film has sensitivity to a little beyond 1.0μm, most aerial infrared film does not go beyond $.9 \mu$m. Filters which cut off the shorter wavelengths of the visible region (blue and violet) are commonly used to enhance the atmospheric penetration capability of the film. In fact, filters are employed often with black and white infrared film to use only the far red and infrared (.7 to $.9 \mu$m) although it is sensitive to the whole visible region. Aside from its haze penetration capability, infrared film with filters is useful in the delineation of water bodies and the detection of distressed plants. Water absorbs infrared radiation more effectively than it does the visible wavelengths so that a land/water boundary or shoreline is usually more sharply defined on infrared than on conventional film. The ability of infrared film to detect differences in the amounts of infrared radiation reflected from healthy and distressed plants has been mentioned in Chapter 1; other uses of this film will be discussed later.

The color version of infrared (IR) film is referred to variously as either color infrared, camouflage detection film, or false-color film. The latter name reflects the fact that most features don't appear in their normal colors on photos made with this film (plate 3). Healthy green vegetation normally is shown in red, for example, and bare soil is greenish blue. Color infrared film, like conventional color film, has three layers of emulsion, each with different dyes and each sensitized to different wavelengths. There is no layer sensitized to blue wavelengths in this film, however; one layer is sensitive to infrared (to .9μm) instead. The greater the amount of reflected infrared radiation received by the film, the greater is the effect upon this layer. Subjects reflecting infrared radiation strongly or in differing amounts are thus identifiable by the tones produced, and again the greater variety of color and tone combinations provides greater flexibility of use than do shades of gray. As in conventional photography, color infrared film presently enjoys greater popularity than black and white. Indeed, new uses for this film are proving it has great versatility of application.

Although infrared photography is commonly considered a part of "photographic remote sensing" because it employs a film/camera sensor system, the term photographic is, strictly speaking, inappropriate. The word "photo" originally meant "of or produced by light", and infrared wavelengths are not within the light region of the spectrum. Usage customarily ignores this distinction, however, partially because most photographic infrared does utilize part of the visible spectrum. Another reason may be that it helps to distinguish remote sensing which uses only infrared energy immediately adjacent to the visible from the sensing that employs energy from other portions of the infrared region.

Sensing the Ultraviolet

As mentioned earlier, developments in remote sensing through ultraviolet or shorter wavelengths have not progressed as rapidly as those which use longer wavelength energy. Chief among the problems encountered is the effect of the atmosphere on these shorter wavelengths. Attenuation, or scattering and absorption by the atmosphere, is the common fate of much shorter wavelength energy. We perceive the sky to be blue because as we look up into the atmosphere we see predominantly blue light, that part of the visible spectrum scattered most by the atmosphere. In fact, certain physical laws indicate that we should not be able to use ultraviolet energy at all if it must pass through any significant amount of

Figure 2-3: An example of ultraviolet imagery. The quality of the picture is misleading in that it appears to have been made by a camera. The UV radiation was collected by a scanner instead and then used to make a photo. (Courtesy of Texas Instruments, Inc.)

atmosphere before reaching a sensor. Ingenuity, though, has made it possible; a device known as a scanner collects a total signal, including a lot of "noise" (undesired radiation), and the direct ultraviolet subsequently is separated electronically from the scattered radiation (fig. 2-3).

An airborne optical mechanical scanner (usually just called a "scanner") collects energy from the landscape below in a series of scanlines, each of which is perpendicular to the line of flight. The energy is received by a rotating mirror and reflected into a small telescope which focuses the energy on the scanner's detector. The rate of rotation of the mirror is adjusted to the velocity of the sensor platform so that, ideally, adjoining scanlines do not overlap or leave gaps in the landscape scene. The energy received by the detector (ultraviolet in this case) varies in signal strength as the elements of the landscape being scanned vary in character. These variations in signal strength, each with an "address" on the scanline, can be stored on tape. Later they can be filtered to remove the noise and

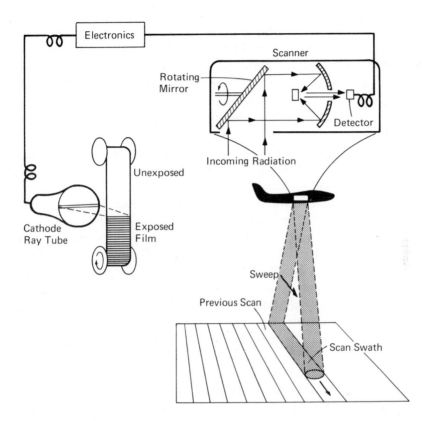

Figure 2-4: The scanner operation principle. Variations in the strength of the signals collected by the scanner are converted to light and dark tones on a cathode ray tube and the result photographed.

played back for display on a TV picture tube as tones of dark or light, depending upon the signal strength. The composite of all the signals, each in place on its scanline, with all of the scanlines adjoining, provides a black and white picture not unlike a photo. In fact, one system synchronizes the exposure of ordinary film to the line-by-line display of the signals on a cathode ray (TV) tube and produces a composite photo (fig. 2-4). Thus do we use visible light to find out what we would see if the eye could utilize ultraviolet light.

Thermal Infrared Imagery

The infrared region includes wavelengths from just beyond .7μm to 1000μm (one millimeter). We have been talking about the shortest of these in discussing photographic infrared, the wavelengths close to 1.0μm. The total infrared region is commonly subdivided into several parts, of which the one called the "near infrared" includes these shortest wavelengths and extends to 3μm. In the following section we will be concerned primarily with energy from the two other parts of the infrared region, the "middle" and "far infrared".

The sun, with a surface temperature near 6000°K, has a radiation peak in the wavelengths of approximately $.4\mu$m to $.7\mu$m, and most of the remainder of its radiation is in the shorter wavelengths of the infrared region. The earth radiates energy also, but its markedly lower surface temperature results in the peak radiation being at longer wavelengths. Earth surface temperatures vary widely, both by location and time, but a figure of 300°K (27°C) is considered an acceptable mean. Peak radiation at this temperature under idealized conditions would occur at 9.7μm. Conditions attendant to earth radiation are not ideal in this theoretical sense, however, and we cannot fix earth radiation as precisely as solar. Additionally, radiation curves are not as sharply peaked at lower temperatures so that the earth's surface radiates substantial amounts of energy over a broader range of wavelengths. We may settle for a general statement that earth radiation is dominantly comprised of wavelengths that are longer than those of peak solar radiation by a factor of between 10 and 20. Non-photographic infrared remote sensing is concerned principally with energy having wavelengths from about 3μm to about 14μm.

Some mental reflection at this point should bring forth the realization that earth radiation overlaps a portion of the solar infrared. During daylight hours, a sensor operating in the 3μm to 5μm band will be receiving both reflected solar infrared radiation and direct earth infrared radiation. But this variety of remote sensing does not require light—indeed it is not concerned with light. In this case, we are seeking to identify differences in emitted thermal radiation using both the middle and far infrared subregions, together termed the "thermal infrared" portion of the infrared spectral region (see fig. 2-1).

Thermal Sensing

The sensor used in this region is called a thermal scanner and it is similar to the scanner described in the section on UV. Two differences are noteworthy here: the nature of the detector and the necessity for temperature control.

Detector elements sensitive to energy wavelengths longer than 3μm make use of materials with strange names like mercury:germanium and indium antimonide. But the problem encountered in attempting to sense thermal radiation typical of the earth's surface is more intriguing. The thermal scanner is supposed to sense variation in emitted thermal radiation caused partly by temperatures similar to those of the scanner during operation. How can we keep the scanner itself from affecting the data? The solution is to cool the detector to extremely low temperatures and then enclose it in a heatproof "box"—just as light-sensitive film in a camera is enclosed in a lightproof box. The detectors of thermal scanners that operate in the 3 to 5μm band are cooled to about $-200°$C with liquid nitrogen. For sensing at longer wavelengths such as 8 to 14μm, temperatures closer to absolute zero are required.

Otherwise, the thermal scanner operates in essentially the same manner as the scanner used in ultraviolet sensing (in fact, some scanner systems are equipped to sense both UV and IR), and images of infrared radiation patterns may be produced similarly. Such photo-like reproductions are called infrared imagery, however, since their basis is the emitted infrared energy received by the scanner rather than reflected solar UV or IR energy. While such imagery does give us information about temperatures at the scene surveyed, we must remember that temperature differences alone do not explain contrasts in signal strength. There are at least three additional considerations. Imagery made during daylight hours in the band of approximately 3 to 4μm may be based on nearly equal amounts of reflected solar infrared energy and radiated earth infrared energy. If we want only the latter, we may either use the scanner at night or operate it at longer wavelengths such as the 8 to 14μm band where the solar component is a minor one. Another consideration is that differences in signal strength are not caused only by temperature differences. There is a factor called emissivity which can vary with, for example, the nature of the material, or the form of the surface. Two subjects with the same temperature can give off different amounts of infrared radiation if their emissivities differ. Then, too, the atmosphere between the subject and the sensor may alter the signal. The user who would obtain temperature information from scanner data must take these factors into consideration.

In discussing the ultraviolet, we referred to the scattering effect which the atmosphere has on short wavelength energy. At the wavelengths we are considering here, a related problem occurs. The earth's atmosphere directly absorbs some of the wavelengths of the infrared region, preventing their transmission and thus limiting their use for remote sensing. There are bands of wavelengths in the infrared, however, which are little affected by absorption. The atmosphere's transparency to the wavelengths of these bands has given rise to the term "windows"; such windows are identified by the wavelengths transmitted. This property is one of the reasons most of the thermal infrared remote sensing to date has been in two bands, from 3.5 to 5.5μm and from 8 to 14μm.

This discussion may seem to imply the existence of a remote sensing capability gap between photographic infrared to .9μm and thermal infrared beginning at about 3.0μm. Actually, an optical mechanical scanner can and does operate at this band. In fact, scanners may be used throughout the visible region also, but for most purposes the superior resolution obtainable with the conventional camera is preferred.

Microwave Sensors

The microwave region of the spectrum includes wavelengths from approximately one millimeter to several meters, and is thus another large spectral region. As in the case of the infrared region, not all parts of it are used; but increasingly the longer wavelengths are the object of experimentation as remote sensing hardware is sophisticated. Most of the effort to date has utilized the shorter wavelength bands of the region.

The two microwave sensors that are treated here are radar and the microwave radiometer. Although they may use some of the same wavelengths of energy, the microwave radiometer is sensing "natural" radiation from the scene being surveyed whereas a radar set generates the energy with which it senses. For this reason, radar is called an active sensor; the microwave radiometer and the other sensors discussed in this chapter are called passive sensors.

Radar

The term "radar" is an acronym created from RAdio Detection And Ranging; its early use was in the detection and location of various types of military targets. The time required for a transmitted signal to reach a target and return and the angular relationships between sender and target were convertible into distance and direction information. One way to make the information usable was to display it on a viewing scope which made the relative locations of the sender and the target evident. Modern imaging radars are based on the same principle, but there are some important differences. Sidelooking airborne radar (SLAR) is the variety of radar that has enjoyed the greatest popularity in remote sensing. As its name implies, SLAR looks to one side of the aircraft, at a right angle to the flight line rather than ahead or in a circular pattern. The returned signals are usually stored on tape and then replayed to create a photo-like image of the scene rather than using a scope display.

The coverage provided by an individual burst of transmitted energy begins not immediately below the aircraft but at some distance away from that point, and extends outward to near the horizon (fig. 2-5). A typical sweep angle totals about 50 degrees. Differences among the elements of the landscape struck by the transmitted energy cause variations in the return signal. The timing of return and angular relationships provide an address along the sweep line for each "bit" of return. As these returns, stored on tape, are later displayed on a cathode ray tube, differences in signal strength appear as light or dark tones. The many "bits" from a given sweep, each at its address and depicted as a bright or dark

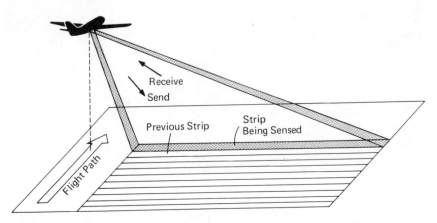

Figure 2-5: Sidelooking airborne radar operation. Signal returns are collected along one side of the flight line at a time as strips of data. Subsequent combination of successive strips provides the basis for a continuous picture.

tone, constitute a line of variable tones. By combining adjacent lines we can create a radar image of the scene which can be photographed, much like a scanner image.

The quality of radar imagery has been so improved that it is sometimes necessary to remember that it is not photography, that it is a different view and should be expected to look different (fig. 2-6). We do not yet understand all that we see on radar imagery, although much research is being and has been devoted to it. Surface shapes are quite important to the return signal, as is the nature of the surface material. Water surfaces usually yield little or no return while industrial areas create a strong return signal. Areas of rugged terrain are depicted by strong tonal contrasts; commercial zones of urban areas generally cause a stronger return than residential areas. Such contrasting responses constitute a basis for some of the current research on experimental applications of radar imagery.

An active sensor such as radar is capable of producing imagery with better resolution than a passive sensor using the same

Figure 2-6: An example of the photographic quality of SLAR. This image shows a volcanic formation on the island of Bali in Indonesia. The aircraft flight line was along the top of the picture. "Shadows" are in areas blocked from receiving the transmitted signal and, thus, from which there was no return signal. (Courtesy Goodyear Aerospace Corp./Aeroservice Corp.)

energy bands. For one thing, radar produces two kinds of locational information, angular data and time (distance) data. But the development of synthetic aperture radar increases the quality potential of radar even more. The radar antenna which receives the return signal may be likened to the lens on a camera: the better the lens, the better the picture. Generally speaking, the longer the radar antenna the greater the resolution possible. But there are obvious limits to the antenna length that an aircraft can accommodate. It was discovered, however, that if certain aspects of the return signal (phase and amplitude data) were utilized correctly, the data could be processed electronically the same way a large receiving antenna would process them. With this procedure a synthetic aperture radar with an antenna only a meter or two in length can perform in a manner equivalent to a radar with a real antenna 600 meters long!

The signal that radar sends out may be polarized, horizontally or vertically. The return signals also exhibit polarization. Imagery may be made from signals with the same polarization as those sent or signals with the opposite polarization, since both types are returned. The imagery of an area produced by cross-polarization (HV) sensing differs from that produced by like-polarization (HH) (fig. 2-7). These differences are believed to have utility for such tasks as signature identification (Morain and Simonett, 1966).

Radars use a variety of wavelengths and thus have varying capabilities. Although all radar may be used day or night and is unaffected by clouds, shorter wavelength radars can detect rain showers. They also depict contrasting vegetation types differently (see fig. 2-7), although longer wavelength radars seem to ignore vegetation and emphasize the surface beneath. It is easy to be enthusiastic about the potential of radar, but it does have shortcomings. Some are merely bothersome, such as the "foldover" of steep slopes which may occur on imagery of rugged terrain. Perhaps more important is the present uncertainty about what the returns mean. Research will alleviate this problem; in the meantime radar will continue to serve where its abilities have been proven, albeit not completely understood.

Radiometry

The microwave radiometer, a passive sensor, measures the amount of energy at selected wavelengths emanating from a scene.

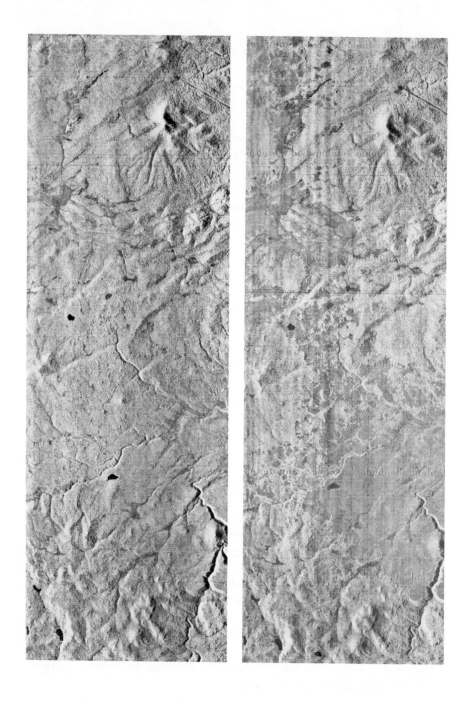

Figure 2-7: A comparison of like-polarized and cross-polarized radar imagery. The flight line was along the left-hand edge of each picture. The image at left was produced by a horizontally polarized emitted signal and a similar return signal (HH). As is evident, the cross-polarized HV image at right depicts some patterns differently than the HH. (Courtesy of Westinghouse Electric Corp.)

When the radiometer is directed toward the ground, the signal received by the antenna includes several components. There is an emittance component which is related to surface temperature, a transmitted component with a subsurface origin, and, during the daytime, a reflected component. The atmosphere between the sensor and the scene also affects the received signal. Variations in the signal received by the antenna result in variations of electrical power in the sensor system; an expression of the amount is stored on tape. The composite signals are referred to as brightness temperatures, although the temperature of the surface is only one component. They may be displayed as numerical readouts, as graphs, or in plan forms such as maps. In the latter case, a color is typically assigned to each brightness temperature (or class), and a crude pictorial image in color results (plate 4). The picture is crude because the individual resolution cells which comprise it are typically rectangles or squares, each of which is representative of the average conditions within a substantial ground area. Recent improvements in this type of sensor are providing substantially better imagery.

Wavelengths this long pose definition problems but they have penetration capabilities that shorter wavelengths do not. Information about subsurface conditions such as soil moisture contrasts is believed to be obtainable with radiometers. Curiously, information concerning additional items as diverse as upper atmosphere temperatures and sea state (degree of roughness) also is obtainable with the microwave radiometer. The point was made earlier that at earth surface environment temperatures, peak radiation is near $10\mu m$; but that a graph curve depicting the amount of radiation at different wavelengths is relatively flat. With this sensor we are using some of the less plentiful radiation some distance from the peak, say in the 1 millimeter to several centimeter range. Sensing passive radiation of such wavelengths poses technical difficulties, but it gives us still another way to examine the environment.

Extending the Utility of Sensors

There are additional varieties of remote sensors, but the camera, the scanner, radar, and the radiometer are the ones that have been most used in remote sensing to date. Thus, we have alternatives to which to turn when the conventional camera proves unequal to the task, and each alternative constitutes a separate though related aspect of remote sensing. There are problems, however, which no one remote sensor can handle alone but which will yield to the combined use of several systems. Since energy from several parts or bands of the electromagnetic spectrum is involved in this sort of operation, it is called "multiband" or "multispectral" analysis. Brief reference was made earlier to the use of several types of imagery in the search for signatures; multispectral analysis offers a greater range of possible signature parameters than those obtainable with only one sensor or from only one spectral band.

Multispectral analysis need not be dependent upon imagery from more than one type of sensor. By using several photographic transparencies, each made with a different wavelength of light, in conjunction with projectors equipped with colored filters we can reconstitute the several images into one in which the strengths of individual colors may be varied at will. The number of combinations possible with such an arrangement is quite large, and the possibilities for finding a combination which make the desired information more evident are increased accordingly. Thus, the utility of much remote sensor imagery is not limited to what is immediately apparent on it; use in combination with other imagery forms or further treatment may make it additionally useful.

A variation on this latter theme is known as enhancement. There are several varieties, but color enhancement will illustrate the principle. Often tonal variations from place to place on a piece of imagery will indicate change to be present, but the change may be so gradual or so subtle that no boundaries are discernible. For example, in an area of mixed vegetation all of the species can be similar enough that they do not cause substantial differences in the signals received by the sensor. Or an area of water may have temperatures that gradually change from place to place with an accompanying continual alteration of the tonal indication of temperature. There are instruments capable of detecting differences in the optical density of a film emulsion that the eye cannot detect, and they can identify the location of boundaries between image classes based on whatever parameters are desired. Taped data can be reviewed electronically in similar fashion. By assigning brightly contrasting colors to the several image classes and reconstituting the imagery with the colors replacing the subtle tonal variations of the original, boundaries become strikingly visible, as do rates of change from class to class (plate 6).

Enhancement involves the reworking of data after it has been received. With some of the data stored on tape, astonishing possibilities exist. Electronic manipulation of the total data can increase or decrease emphasis, extract data selectively from the total, test combinations of signature parameters, and examine data characteristics that would not show up on imagery. Indeed, discussion of this topic after the foregoing listing of capabilities of remote sensing is apt to cause the reader to forget the admonitions of Chapter 1. They will not be repeated here; only a reminder is offered. In this chapter and the several which follow, an emphasis on the positive seems reasonable despite the reality of limitations.

Chapter   3

Remote Sensing of the Natural Environment

The natural environment is much more to man than the stage on which he plays out his various roles; it is also his resource base. From it he obtains the necessities of life as well as those luxuries his particular circumstance allows. For some men, luxuries have come to be commonplace, and nature seems to be limitlessly bounteous, but such is not the case for most. There are many reasons for differences in men's circumstances, but foremost among them is lack of knowledge. We must know how to use resources in order to derive benefit from them, but a knowledge of what exists is perhaps most basic. Our knowledge of our resource base has many (and some are large) blank spaces, and parts of it include misinformation, either originally incorrect or now outdated. The world circumstances we face now and in the immediate future require a better inventory.

Change is a reality of nature, although its temporal and spatial disguises are misleading. Alterations which occur either so slowly as to be imperceptible or in places so widely scattered as to go unnoticed can, in time, substantially transform an area. Some natural change is not so gradual, of course, and is less likely to go unnoticed, although its long-term effect may be no more obvious. Among the factors causing changes in nature, man has come to occupy a prominent position. Perhaps it is chiefly because of man's activities that we have realized that inventory alone is an inadequate basis for resource appraisal. With some resources, it is just as important to monitor changes.

The natural environment is not always a benign supporter of man's activities. In ways which vary from the abrupt violence of the tornado or the earthquake to the insidious discouragement of agricultural drought, forces of the environment sometimes seem bent on man's destruction. We have come a long way since the belief that volcanic eruption was a god's expression of anger or thunder the sound of trolls bowling, but we have much farther to go. We can make it rain—sometimes—but we cannot control hurricanes. We know how earthquakes occur, but not how to predict them. Perhaps some of these paroxysms of nature will never be controllable; it is certain they never will be unless we learn to understand them. To be able to lessen their effect or even to predict them accurately is a worthy enough goal—witness the reduction in loss of life from hurricanes that improved prediction has made possible. For either prediction or modification, a more complete understanding of the environment offers the best possibility. Such understanding is dependent upon improved observation. To be sure, modeling in the laboratory has provided some exciting results, but the models must be based on a knowledge of what occurs in nature.

To inventory, monitor, and understand the natural environment adequately we need more and better ways to collect information. Remote sensing offers some new—and sometimes better—ways. In the sections which follow, examples of application are provided for each of several aspects of the natural environment: the *lithosphere*, the solid rocky earth itself; the *atmosphere* or the envelope of gases surrounding the earth; the *hydrosphere* or the water bodies of the earth; and the *biosphere* or the thin skin of life on or near the surface.

The Lithosphere

Students of the lithosphere include, among others, geologists and geomorphologists; the focus of their attention is the rocky crust of the earth and its veneer of weathered rock. Some study these things out of intellectual curiosity or to advance scientific knowledge. Others have more pragmatic goals, and foremost among these is the endless search for mineral resources, from sand and gravel to gold and silver. Individual interests include such considerations as rock structures, surface form, and the nature of the rocks and minerals which comprise the solid earth. These topics have been studied for years by various methods, including conventional photography, but remote sensing is providing some new methods for problem solving and some insights that are altering our picture of the lithosphere. Additionally, the possibility of satellite imagery being useful for some of these studies has especial appeal because of its potential for reducing an enormous task to more manageable proportions.

There is a certain chicken-and-egg aspect to studies of landforms, rock structures, or earth materials in that each usually can tell you something about the others. One reason for studying surface form is that it is often indicative of the structures which underlie an area. In turn, some structures and some rock types are suggestive of the possibilities for mineral deposits; the association between petroleum and anticlinal or domed structures is an oft-cited example. Yet structures are one of the causes of surface form, as are earth materials.

We lack good maps of surface form in many parts of the world. Topographic maps formerly were made by ground survey, a laborious process which required tramping over the area with surveying instruments. Sometime ago we began switching to the use of aerial photos; with a few ground-checked reference points, stereo photo coverage provides the basis for map making in the laboratory with photointerpretation instruments. This quicker, cheaper, and better method appeals to younger or less developed nations with large mapping efforts in prospect. But the task of obtaining good airphoto coverage has its own problems, one of which is cloudy weather. It may not seem that this factor should be so much of an obstacle, but it can be. Some parts of the world are seldom cloud-free, and matching all the pre-mission preparations with the erratic occurrence of acceptable weather can be frustrating and expensive. Obviously this problem would be eliminated if adequate surface form imagery could be obtained with radar. Its capability for portraying surface irregularity has been mentioned. In a much publicized experiment several years ago, SLAR imagery revealed several errors in the most up-to-date maps of a portion of cloud-prone Panama, including a mislocated major stream and an incorrectly depicted upland area. (MacDonald, 1969).

The cloudiness of the humid tropics presents a problem for aerial photography; yet these regions have great areal extent and are imperfectly known. Such basic developmental efforts as establishing transportation arteries would be aided by improved landform maps. Surface form depiction on radar offers other information also, however. Different types of rocks weather and erode differently, creating contrasting signatures on the imagery. Structural features such as faults or crustal fracture zones also are often evident from radar imagery that emphasizes them despite the dense vegetation cover typical of such areas. The information on structure and lithology, often little known in these areas, may be as valuable as that on surface form.

The term rock structure refers to the shapes of rock masses, the directional attitudes of these masses, fractures of the crust, and the like. Structures have a great range of sizes. Swarms of small cracks created by shifts in the crust may cover only a few acres; a continuous zone of folded rocks such as the Appalachian Mountains may extend across many states; the great tectonic plates of the crust which are related to continental drift boggle the mind with their size.

Study of the larger structural features of the globe puts man in a position comparable to an ant on a Persian rug seeking to visualize the rug's pattern. Before satellite imagery, we sought to tie together bits and pieces of small area studies on maps, or to make a mosaic from many photos and reduce it photographically in order to see the larger picture. Neither solution was entirely satisfactory. What went on the map was dependent first upon the accuracy of the individual studies made by many different people, and second, the judgment of the map maker who selected the information to be used from the studies. The individual photos which went into the mosaic were taken at different times and developed at different times so that tonal variation, so important to interpretation, became an unknown variable.

A photo taken by an astronaut or returned from a satellite shows thousands of square miles on one frame, all parts of which were imaged at the same instant under the same conditions. To be sure, the creek that runs across your property and the ridge to the west of it may have disappeared completely at such scales, but the location of a great fault extending across many states is visible because the discontinuous pieces of the line now can be seen clearly to belong to the same feature. Some features, seemingly unimportant or simply puzzling when viewed in a limited context, fall into place on a total view. Study of an Apollo 9 photo by one scientist revealed information, undetected or misinterpreted by earlier studies with conventional methods, which led to a new hypothesis of plate tectonics in the Red Sea region.

The lower resolution of much space imagery does limit its utility for many tasks, but intuitive association with larger features that are visible and maximum use of tonal or hue variations and other signature components make possible the extraction of more information than is immediately apparent. Distinction among gross rock types and the identification of major lithic regions has been achieved on some space-derived imagery (plate 5). Even information that is this highly generalized is valuable, for our present maps of some parts of the world either are even more generalized or are non-existent. Interestingly, the repetitive nature of some satellite imagery affords unexpected dividends to offset low resolution. Changes in the seasonal appearance of landscapes can become a factor. Snow patterns are a noteworthy example in that they often emphasize features of drainage patterns (fig. 3-1), which can be revealing indicators of structure, surface form, and rock type.

Remote sensors which take a closer look at the surface have been applied to rock type and structure studies. The microwave radiometer has detected differences in brightness temperature among some types of rocks, and the thermal scanner has been used in studies of structure as well as lithology. These instruments have

Figure 3-1: A view of Alaskan snow cover from Nimbus 4. This photo represents a rare event, a day on which the entire state of Alaska was cloud free. The white on the land is snow and the fretwork of dark lines in it is the drainage system outline. (NASA photo)

identified fault zones or other cracks in bedrock by detecting differences in emitted radiation caused by the presence of moisture in the cracks. Sometimes this difference will manifest itself through a layer of weathered rock and a cover of vegetation, both of which would mask the feature to the eye or to the camera.

One of the more striking examples of making visible the invisible appeared in a study (Sabins, 1969) which emphasized the importance of time of day to thermal scanning. In a section of the California desert being imaged experimentally with a thermal infrared scanner, conventional photography revealed nothing and the author of the study believes he would have seen nothing significant had he walked across the area (fig. 3-2). Daytime scanner imagery was equally unrevealing, but pre-dawn thermal image revealed both lithologic and structural features. Most notable was an eroded fold, an anticline, with a fault passing through it. The differences in emitted radiation from the various rock layers during the day were too small to distinguish them, whereas differential cooling rates at night made it possible to do so. Several studies have indicated similar findings in different situations; it appears that the early morning hours are especially good for detecting some types of subtle differences.

It should be noted that the identification of the structure in this example came as a by-product of being able to see the shapes of the rock layers as revealed by their radiation emittance differences. The structure itself tells us a little about the rock type; the layered pattern suggests the rocks are sedimentary although what kind of sedimentary rocks they are remains a question. Surface form was of no value in this circumstance since the author of the study indicates there was no surface expression of the feature. This fact plus the indicated similarity of the several rock layers, with perhaps a little help from a veneer of sand, made it invisible to the eye and the camera.

Research to establish radiometric and other signature parameters which will pin down rock types more closely goes on, but such identification is a difficult task. For one thing, with thermal imagery the signature will vary with the time of day, and two different rock types may produce the same signature at different times. Then too, there are so many types of rock, many of which differ from others so little that their precise identification by remote sensing is unlikely. Still, we can make gross distinctions now; and signature research, perhaps through multiband sensing, will increase the possibilities for narrowing identification.

The several capabilities for mapping lithology, form, or

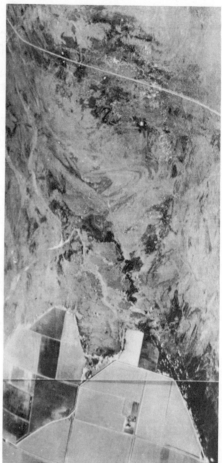

Figure 3-2: Comparison of thermal IR imagery and conventional photo of an irrigated area in California. Structure revealed on the thermal imagery (right) which is completely missed by the camera includes a small fold and a fault. (Courtesy of Floyd Sabins and the Geological Society of America)

structure at various scales from airborne or space platforms, by day or night, and in inclement weather are promises by remote sensing that could revolutionize our knowledge of the lithosphere. If they materialize, they could inaugurate an age of discovery that would rival the 16th century in its importance to mankind.

The Atmosphere

Most of us do not think of the atmosphere as one of our natural resources, although it provides the most immediate necessity for life, air to breathe. Another necessity for land life, fresh water, may have its origin in the sea, but it is the atmosphere that makes it available through the hydrologic cycle. The growing problem of air pollution and the rapidly diminishing gap between water supplies and water need are changing our conception of the atmosphere's importance. Additionally, increasing unwillingness to endure the excesses of diverse atmospheric moods has prompted investigations of possible recourse, and in a few cases there are definite indications that some degree of control could be achieved.

The task of trying to sample and monitor anything as vast and inconstant as the ocean of air that surrounds us is monumental. Even if we concern ourselves primarily with only the bottom 10 to 20 miles where weather occurs, the problems are limitless; and the closer to the earth's surface, the greater is the atmosphere's variability. At the surface, for example, temperatures differ on the opposite sides of a hill or between the suburbs and downtown. For years we depended upon data collected at weather stations and by ships, aircraft, and a few balloons. Plotted on a small-scale world map, these atmospheric sampling points look impressive; today some 25,000 surface observations are made daily. At best, though, perhaps 20 percent of the globe is adequately covered (Roberts, 1972). Differences in energy received and lost in the tropics and at the poles is a basic cause for atmospheric circulation, yet these two areas are the least well known. The oceans, which cover nearly three-fourths of the globe, are monitored by a fraction of the number of stations on land; the southern hemisphere oceans especially have great areas with data gaps. To be sure, since the variation from place to place in the atmosphere over the oceans is more gradual than on land, the data from a given point is likely to be representative for a larger area. But there is so much ocean that small changes are magnified in importance. Moreover, the atmosphere is so mobile that change can occur quite abruptly.

One solution to increasing the coverage of the globe is the use of automated unmanned weather stations, some of which are in service both on land and at sea. It is unrealistic to contemplate an infinite number of weather stations though; mobile instrument clusters which can survey all areas repeatedly are preferable. What is needed is a continuous scanning of the total surface to provide areal data to complement the point data from ground stations. If the instrument system is complementary also in providing a view from above to go with the one from below, so much the better. The weather satellite satisfies these requirements.

Weather Satellites

The United States has launched several families of weather satellites. Perhaps Explorer VI, launched in 1959, should be called the first since it did return televised cloud cover pictures, but the TIROS series are usually thought of as being our first successful weather satellites. The acronym stands for Television InfraRed Observation System. TIROS I was launched in 1960, the first of ten in the series. The sensors varied on these satellites as we experimented with systems that would return televised cloud pattern pictures and both reflected and emitted infrared radiation data. Some of the information returned was used operationally by the National Weather Service, but the TIROS satellites also were paving the way for a wholly operational series, the ESSA (Environmental Science Services Administration) satellites.

ESSA I and II were launched early in 1966, shortly before TIROS X ended operation. The ESSA series also returned televised pictures of the earth's surface and cloud patterns as well as infrared data. Two kinds of camera systems were used, however, in alternate ESSA satellites. The Automatic Picture Transmission (APT) camera transmits continuously the scene it is viewing, and the transmission can be received by anyone with the proper equipment. Many have taken advantage of this accessibility—transmissions have been received in more than 90 countries and territories. The necessary receiver is remarkably inexpensive and at least one group of high school students has constructed one. The other system, the Advanced Vidicon Camera System (AVCS), stores the data it collects and transmits it only on command to a few stations. Satellites equipped with these systems have been positioned several hundred miles above the earth in orbits which take them from pole to pole so that combining successive orbit transmissions can provide complete global coverage. The imagery has rather low resolution, on the order of two or three miles, so that its utility for the study of surface features is limited (fig. 3-3). A surprising variety of non-meteorological studies have made use of the imagery, however, mostly to investigate larger features of the environment (Sabatini et al., 1971).

The infrared data provides information for several purposes, but one of the best known uses is for the determination of cloud types. Cloud top temperatures are related to cloud heights which in turn help identify types. Cloud types indicate atmospheric characteristics, and monitoring cloud development provides the meteorologist information on the potential for weather changes.

The Nimbus research and development satellites (fig. 3-4) are forebears of future generations of satellites. They carry a much more impressive battery of sensors than either TIROS or ESSA. Nimbus IV, which was launched in 1970, has ten. In addition to an improved television camera, several varieties of infrared sensors are on board. The Temperature Humidity Infrared Radiometer (THIR), for example, provides cloud top and earth surface equivalent temperature data (fig. 3-5), as well as water vapor content data for the upper atmosphere. Several other instruments sense temperature indicators and water vapor content at various levels, providing the basis for constructing an atmospheric profile, an important element in forecasting. The IRLS (Interrogation, Recording, and Location System) is an especially interesting device. It interrogates automatic sensor systems at the surface or in the air (balloons) for data they have collected, and retransmits the data to ground stations.

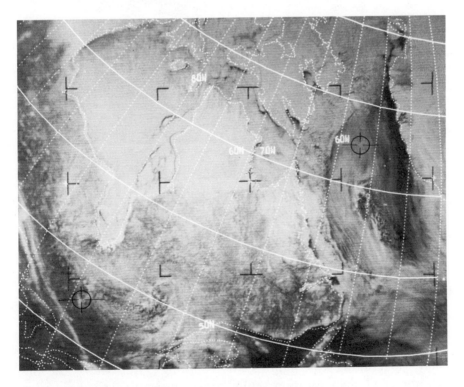

Figure 3-3: ESSA weather satellite AVCS image of northeastern Canada. The west edge of Greenland is at upper right, Hudsons Bay is at upper left, and part of Newfoundland is in lower right. (NASA photo)

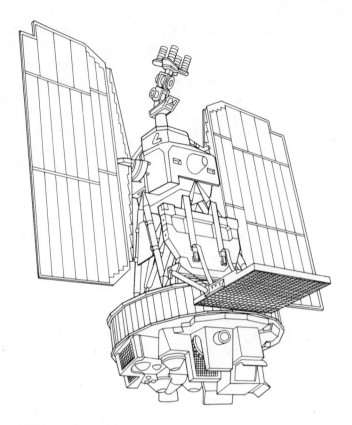

Figure 3-4: The Nimbus 5 spacecraft. The sensor cluster is at the base and the two paddle-like solar array panels which collect solar energy to power the equipment are on either side.

The counterpart of the automated weather station floating in mid-ocean is the atmospheric sensor package mounted in a constant-level balloon, drifting along through the upper atmosphere. We have long made use of instrument packages called radiosondes, borne aloft by large balloons, which transmit data that provide a vertical profile of temperature, moisture, and pressure along the path of the rising balloon. These newer devices, called GHOSTs (for Global HOrizontal Sounding Technique), stop rising at a selected altitude, predetermined by the amount of helium put into the balloon and limitations on the balloon's expansion. They may float along for months, transmitting information on the weather and their location; comparing consecutive positions also provides data on upper-air wind speed and direction.

Another type of weather satellite is NASA's Application Technology Satellite (ATS). ATS III provides a good example of

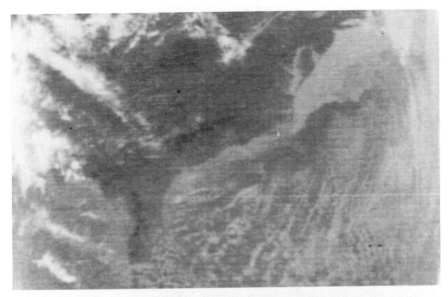

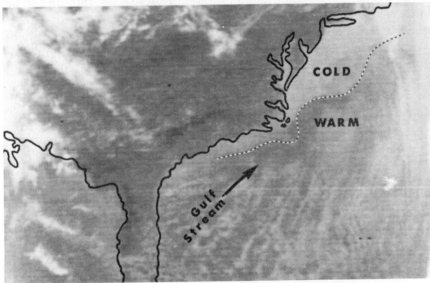

Figure 3-5: The Gulf Stream margin along the U.S. east coast. In this thermal image obtained with the THIR sensor on Nimbus 4, cooler temperatures are light-toned, warmer are dark. (NASA photo)

truly *remote* sensing; the satellite is at an altitude of more than 22,000 miles above the earth. Most weather satellites are designed to operate several hundred miles high and to orbit the earth about once every 90 to 120 minutes. ATS III is 22,300 miles away and not farther nor closer for a special reason. In general, the farther from earth a satellite is located, the longer it requires to orbit the earth. Because of its altitude and its position over the equator, ATS III has an orbiting velocity that exactly matches the speed of earth's rotation. The result is that it hovers suspended over the same location until it is moved on command, which occasionally it is, for orbit correction. But more interesting are the movements which give it greater flexibility of use. Recently ATS III was moved from a position used to monitor severe storms in the mid-United States in early summer to a position farther east to watch for hurricane development in the fall.

ATS III has three vidicon cameras, each of which senses a different wavelength of light. The signals transmitted to earth provide the basis for creating a color image of the earth and its cloud patterns (plate 7). From its altitude, ATS can see virtually the full earth disc and is capable of creating a complete image every 20 minutes or so. By combining a number of these stills, motion can be imparted to the view just as cartoon characters are animated. Hence, the majestic convolutions of great weather systems half a continent wide can be observed in motion and their developmental status can be more clearly perceived. Although it may seem unlikely that much could be seen from a distance of 22,300 miles, major cloud patterns are quite visible and very revealing. The position of jet streams, the incubation of hurricanes, and the strengthening or weakening of mid-latitude cyclones are indicated by cloud pattern and type information. Data from the newest of these earth-synchronous satellites, GOES (for Geostationary Operational Environmental Satellite), may even be able to provide timely warning of possible tornado developments. GOES also has the interrogation capability pioneered by Nimbus IV.

The successor to the ESSA series is the NOAA satellite, named for the National Oceanic and Atmospheric Administration (which includes the National Weather Service). NOAA satellites, benefiting from NASA's experimental Nimbus series, are much more sophisticated than the last ESSA satellites. Their sensor package includes an infrared system that measures earth heat radiation, an infrared scanner which can provide day or night cloud and surface imagery, improved television cameras with both AVCS and APT capability, and a sensor which monitors solar storms.

The volume of data the satellites have furnished is about as incomprehensible as the national debt. TIROS I alone returned some 23,000 cloud photos and TIROS II over 35,000. These are only two of ten satellites in one series and five series have been named, each with increasing capabilities. Moreover, the data potential is even greater than that received, inasmuch as some satellites only operate on command. Since this data is returned as electronic signals with many thousands of "bits" for a single picture, it is evident that considerable amounts of tape storage space would be used if all of the data were stored. GOES is reputed to have the potential capability of returning one hundred billion bits of information in 24 hours (Roberts, 1972). Much of the data from weather satellites is converted into pictures, some reproduced as the information comes in, and stored in that form if at all. But one of the goals of increased data coverage is to employ it in numerical forecasting, with the assistance of computers. Within limits, the greater the volume and variety of information that goes into a forecast, the better it should be. But weather data is perishable—its utility has time limits. We are long past the point where we can deal with the amounts of data required by modern forecasting without computer help. The great speed and memory capacities of today's computers allow them to make calculations, subject many results to comparative analysis, and draw the most likely conclusions from the data far more effectively than could be done otherwise. Some weather maps already are produced completely by automation. Satellite-generated data go from receiver to computer directly and the computer conclusions are fed into another machine that draws a weather map. It is clear that we will need computers with many times the capacity of today's best to handle the floods of data that are imminent.

The increasing application of remote sensing techniques to the study of the atmosphere (fig. 3-6) should produce remarkable results. The extent of data coverage that satellite sensors are designed to provide is impressive and includes: continuous imagery of the surface and cloud patterns from television camera systems with both stored and direct readout capability; real-time infrared data which can be displayed on a photofacsimile recorder as thermal imagery for day or night coverage; several thermal and humidity sensors for determining water vapor and temperature characteristics of the lower and upper atmosphere as well as at selected levels to provide atmospheric profile data; a solar flux monitor to observe variations in certain types of solar energy believed to affect our atmosphere; and an ultraviolet spectrometer to monitor the ozone distribution of the upper air for global energy budget studies. And such a listing does not include other applications of satellite information such as snow cover extent (see fig. 3-1). The transformation of half

Figure 3-6: A composite view of North America made with the High Resolution InfraRed (HRIR) sensor on Nimbus 3. Notice the spiral cloud pattern of the cyclonic storm northwest of the Great Lakes. (NASA photo)

of the United States with the passage of a strong winter snowstorm is impressive to see, but more important is the effect that snow cover will have on the modification of air masses that subsequently will pass over it, and the significance to atmospheric energy budget calculations for numerical weather forecasting.

Add to the satellite information the data from GHOSTs, from weather buoys, and from conventional stations on land, sea, and in the air, and it may seem the millennium in meteorology has arrived. Not yet, though—satellites still malfunction, as do their sensors. Some sensors are still experimental rather than operational, and we are not certain how to interpret some of the data we receive. Also, since we lack computers and other hardware capable of handling the masses of data theoretically obtainable, for many tasks we will be dependent upon more conventional data sources for some years yet. Indeed, the abandonment of such sources is neither foreseen nor sought.

The World Meteorological Organization (WMO) and its special programs will become more rather than less important in the years ahead. The World Weather Program, with its operational phase, World Weather Watch, and the research-oriented Global Atmospheric Research Program are examples of efforts to use the new data. Local forecasts would not seem to require all this effort, but, in fact, even local forecasts of extended duration do benefit from some of it. Moreover, local forecasts alone have long been an inadequate view of weather; transport, agriculture, and economics have world dimensions—as does disaster. Included among the capabilities of some of the new atmospheric sensors is the detection and measurement of atmospheric pollutants. It is time that we began to have a more accurate knowledge of what we are doing to the atmosphere, that most basic of our resources.

The Hydrosphere

The term hydrosphere has been used to refer to the water in the atmosphere, but more commonly it serves as a collective term for the earth's waters on the land and in the sea. Probably no facet of environment illustrates better the interrelatedness of the sciences; studies in hydrology or oceanography also may be of much value to meteorologists or geomorphologists. The application of remote sensing techniques to the study of water or water-related problems runs the gamut of recognized disciplines.

Although the quantity of water available always has been of concern to people in the drier areas on the world, today it is commonly being monitored and forecast in humid areas also. Stream flow measurements, ground water level indications, and the depth and areal extent of standing water bodies help provide estimates of water availability. Water in natural storage as snow or ice on the land is also an important supply component, especially in mountainous regions or more level areas with a substantial duration of snow cover. Some of the best storage areas are difficult to reach and have physical characteristics which pose problems for accurate estimation of the water present as snow or ice.

Several experiments with airborne sensor systems have produced encouraging results. Passive microwave brightness temperatures have been found to have direct relationship to the water equivalent of dry snow, and radar has proven capable of mapping old snow deposits through a layer of new snow. Both sensor systems can provide areal data to complement surface sampling, and can thus improve water supply forecasts. Such information can help estimate the potential for either flood or drought. More accurate determination of changes in the size of standing water bodies is provided by the improved delineation of water boundaries afforded by infrared film; such changes also may be valuable as indicators of ground water levels. A problem that often plagues reservoirs and their dams is seepage. Color infrared film highlights the weedy vegetation which establishes itself along seepage drainage lines, and in dry regions sequential photography points it up sharply. In humid regions where vegetation cover is normally continuous, thermal scanner imagery has located seep zones by the differences in radiation which the presence of water causes. Differences in water temperature have also provided the basis for locating springs which feed streams, some of which may be identified on thermal imagery by the plume of different-toned water surfacing and spreading downstream. Studies of the effects streams may have upon lakes into which they flow benefit, too, from knowledge of the temperature differences between them (fig. 3-7).

Students of river mechanics have found color IR film especially useful in the study of sediment transport processes. Varying amounts of suspended sediment produce different water hues on color IR film, and current flow patterns and sediment sources have been identified by using this characteristic. A special type of color film has shown unusual depth penetration capability and has proven valuable for mapping and monitoring changes in submerged offshore features. Fresh-water springs emerging underwater from porous volcanic rocks along the seacoast have been identified by remote sensors, indicating potential well sites on a water-scarce island in the Hawaiian chain. Also, passive microwave radiometry has detected differences in salinity of the water in estuarine areas where no water temperature difference was discernible. Seasonal river discharge variations, coastal and river water mixing, and similar kinds of information are valuable to the fisheries industry, since many shellfish and crustaceans inhabit these interface areas and are affected by such factors.

An earlier section of this chapter discussed numerical weather forecasting, in which larger amounts of data can be incorporated into forecast deliberations through the use of computers. An important component, that of energy-mass budget calculations, seeks to approximate a closed system in order to visualize the situation more accurately. Mathematical models are devised which seek to include the major elements which influence weather conditions and to express them quantitatively. The model relationships are then incorporated into a computer program which can employ current data in the solution of forecasting problems. The computer's capacity enables us to use more data and the computer program causes the relative effects of the several elements to be expressed more concisely; the combination of factors should improve forecast accuracy.

Where large areas are involved, snow cover or soil moisture become important considerations, and some of the more ambitious models would incorporate such data into the calculations. Radiometer and scanner data provides some information on soil conditions and weather satellite imagery depicts snow cover. Even the extent of ice cover on lakes is important to energy budget studies where lakes are numerous or large. A knowledge of the amount of open water is especially important to such studies in the Arctic.

The task of collecting information in the Arctic is challenging. Aside from the obvious difficulties posed by extremes of temperature, there are other natural obstacles. The long polar night approaches six months in the vicinity of the Pole itself, thus hampering or prohibiting visual observation and imaging with light-activated sensors. The ice cover causes problems for surface sensing systems; it is an unreliable surface and has surprising mobility. The ice is relatively thin, averaging three meters, and is subject to stresses which alternately may cause compression ridges or openings. Pieces of ice have been tracked over erratic paths of movement at rates of 50 kilometers per day. Several categories of polar ice have been identified by age differences, which are associated with differences in activity and physical character.

With weather satellite imagery, we are provided with continuous sequential coverage, and a new image of the Arctic is beginning to take shape. Leads, or linear openings in the ice, and polynyas, larger irregularly shaped openings, appear and disappear with greater frequency and in greater numbers than previously was

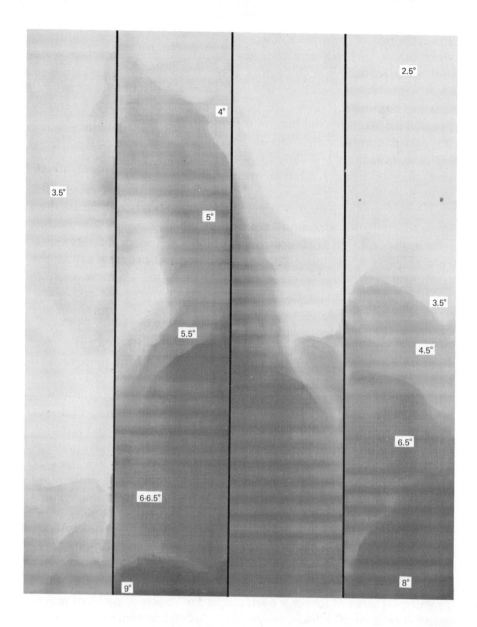

Figure 3-7: Thermal IR depiction of the Niagara River plume. The warmer waters (darker tones) of the Niagara River entering colder Lake Onatario. Studies of physical and biological processes in the lake benefit from such information. Temperatures are in °c. (Courtesy of the Bendix Corp., Ann Arbor)

suspected. Preliminary findings indicate that mean areal ice cover openings are approximately 10 percent of the total area during summer melt. Since it is suspected that a majority of the mass and energy transfer in the Arctic is associated with these openings, monitoring them takes on added significance. Some of these changes occur with the passage of storms over the area, however, and bad weather increases observation difficulties. Radar, which in essence is not weather dependent, has demonstrated its ability to portray ice openings and to detect ice age differences (fig. 3-8). Passive microwave and thermal scanner imagery also can provide ice surface information; all three of these sensor systems are independent of the need for light. Projects such as AIDJEX (Arctic Ice Dynamics Joint EXperiment), inaugurated in 1970 by American and Canadian scientists, are indicative of the increased stature that studies of the Arctic have achieved.

Sea state information is useful to mariners whose progress may be affected by the size and activity of the waves in waters through which they must pass. This information is valuable to the meteorologist also as an indicator of the velocity and direction of the winds causing the waves. Passive microwave brightness temperatures vary with contrasting sea states, and a special type of radar (scatterometer) yields markedly different returns from smooth and disturbed sea surfaces. An interesting association between varying returns from sun glint areas at sea and surface roughness suggests that this phenomenon, usually considered to be undesirable, also may have utility for sea state assessment.

Sea surface temperature data finds application in diverse situations aside from its utility to weather studies. Changes in the position and activity of large surface ocean currents such as the Gulf Stream can be observed on thermal imagery from weather satellites. Oceanographers correlate observed changes in current activity with surface sensor and submerged sensor data to understand better the relationships between the currents and continental shelf and slope waters. Experiments have shown that sea surface temperatures can be obtained by remote sensing with a high degree of accuracy. Airborne radiometers have obtained surface temperature data within .01°C accuracy limits, satellite radiometers to within a few degrees. Associations between concentrations of some varieties of fish and water temperature characteristics have been observed. Establishing this kind of association precisely, plus having ocean surface temperature data routinely available, would obviously have a great impact on the world's fisheries. At the same time, however, conser-

Figure 3-8: Sidelooking airborne radar image of ice floes. The black areas are open water; the differences in tone on the ice are due to surface features, structures within the ice, and differences in the character of the ice. (Courtesy of Westinghouse Electric Corp.)

vation considerations raise sobering questions about having such information.

Encouraging results with microwave, thermal, and ultraviolet sensors in the detection of oil pollution at sea imply the possibility of greater control over this problem in the future. The U.S. Water Quality Administration in 1969 estimated that there are approximately 7,000 oil spills in United States waters per year. While the situation in this country may be atypical, there are enough other nations whose involvement with marine petroleum transport is proportionately comparable that the world dimension of the problem is evident. Since much of the waste petroleum that exists at sea is not the result of accidental spills, the possibility of being able to monitor ships at sea is especially appealing. To be able to do so at night as well as by day and in cloudy weather as well as clear would make such monitoring truly effective. What sensor platforms to use, how to establish a monitor system, and many similar questions are not yet answered finally; but remote sensing provides the ability to see at night and through clouds, and, with satellites, to watch continuously.

The Biosphere

The layer of life that exists at or near the earth's surface includes much that is not considered part of the natural environment.

Evidence of the influence of man is to be found in the flora and fauna of all but the most remote places, and his transformation of densely inhabited areas is abundantly clear. As with so many things, though, a clear-cut distinction between natural and altered biota (plant and animal life) is elusive. It is not necessary to make this distinction here, but admission of the overlap in this treatment does seem appropriate. Most of the discussion of efforts to apply remote sensing techniques to the investigation of life forms has been reserved for the chapter following. Agriculture is clearly a part of the human rather than the natural environment of an area in that it is carried on by man and for man's benefit. Forests and rangelands, however, reflect varying degrees of alteration and management, depending upon the part of the world being considered. The brief glimpse of remote sensing of the biosphere offered here is restricted to a few examples of broad-gauge efforts to investigate items that are clearly natural features; a closer look at other biota applications is found in Chapter 4.

Forests are estimated to cover about one-third of the land surface of the globe. The point has been made earlier that our knowledge of this resource is quite inadequate; we do not even know with accuracy where the forest boundaries are. The Earth Resources Technology Satellites (ERTS) should help us to correct this situation. Experimental mapping with Apollo imagery has shown that forested land can be distinguished from other land uses on space imagery with considerable accuracy. In one study, timing and use of the proper film allowed subdivision of the forested land into three categories: broadleafed forest, needleleaf, and mixed stands. The film was color infrared and the time of year was March, before the deciduous trees had leafed out. Unsophisticated though such information must seem, it is information we do not have for large areas of the world. The immensity of the task of mapping the world's forests makes the possibility of using satellite imagery very appealing. Although the resolution of present ERTS imagery cannot match that of some photographic camera imagery from space, that which has come back from ERTS-1 at this writing is superior to the expectations for it. As a tradeoff for resolution, the ERTS imagery is providing repetitive coverage, which Skylab, for example, does not. Numerous experiments have shown that sequential imagery provides many clues for identification. Known phenological (climatic/growth relationships) characteristics, for example, when combined with visual changes on the imagery as the seasons progress, may prove to be a better basis for identification than superior resolution.

The rangelands of the world, suitable for grazing by domesticated or wild animals but not presently adaptable for crops, are another inadequately known resource. We are aware of changes in the character of this resource and, in some parts of the world, every few years monitor its use with sample ground checks and aerial photography. The sparseness of the vegetation of much rangeland limits its carrying capacity so much, however, that greater expenditures for monitoring seem unwarranted, despite the fact that just such marginal areas are all the more susceptible to irreparable damage.

The possibility of using earth resources satellite data to inventory and monitor the rangelands offers a whole new prospect for future range management. Relatively little additional expenditure should be required; the satellites' sensors automatically include these lands in their coverage. It may seem that little or nothing useful would be obtainable from satellite imagery, since the plants of dry regions are small and scattered. Yet range researchers have found that seasonal changes in plant cover and the emphasis afforded by color infrared allows them to see more than resolution figures would predict. Areal variations of tonal strength in the faint blush of spring plant vigor provide indications of varying amounts of plant cover, suggesting where selected sample data should be obtained by aircraft and ground checks for a better estimate of the resource than we now have. Even in areas where detailed investigations are impractical, mapping of gross differences in cover can provide the basis for monitoring change in subsequent years. Percent cover, in fact, has assumed additional stature as a statistic as a result of research on grasslands. A measurable relationship has been identified between ground plant biomass and percent cover within a given grass type, so that where grass type is established, percent cover maps may be converted into standing crop biomass maps. This relationship is expected to figure importantly in the presently ongoing United States Grassland Biome Program, a part of the International Biological Program.

Finally, despite their limited extent to date, efforts to apply remote sensing to wildlife investigations deserve some mention. Counts of migratory birds at flyway resting stops are made both easier and more accurate by photographic sampling. Fish spawning grounds have been the object of sampling experiments with various film-filter combinations. Pond counts made by automated scanning devices have been used to set hunting quotas. Thermal infrared night surveys of animals have been attempted, and some wild animals have even been fitted with monitoring devices that send signals capable of detection by orbiting satellites. No form of wildlife is too insignificant to escape the attention of some remote sensing enthusiast: unusually dense mosquito swarms are believed to be responsible for certain types of radar "angels," and attempts are underway to monitor locust swarm breeding areas from satellites.

Chapter 4

Remote Sensing of
Man's Use of the Land

Man is distributed very unevenly over the earth. In spite of his numbers and the fact that less than 30 percent of the earth's surface is land, there still are vast areas which lack permanent settlement. As the only human inhabitants of Antarctica, scientists are sojourners whose time there is brief. Although some nomads still utilize parts of the Sahara, there are sections of that enormous desert that even they avoid. The opposite extreme is all too familiar for most of us. Not content with occupying every square foot of available surface space in certain favored locations, we pile ourselves many stories high so that we are literally on top of one another in urban areas.

Differences in population density are often a reflection of differences in land use. Some types of land use, such as industry, require very little space and may be associated with great density of population. Others—agriculture, for example—require greater amounts of space. Such things as variability in levels of living and associated technological capabilities make for great regional contrasts.

For most of us in North America, there is little real awareness of the varying space requirements of different land uses. Our ability to move from one urban center to another with increased rapidity has relegated the space between them to a nuisance category, and the incredible efficiency of our workers in primary production activities such as agriculture has reduced their numbers to such modest proportions that space requirements, other than those in our immediate vicinity, occupy a minor place in our minds.

There are some, though, for whom land use is a matter of continuous and growing concern. Although they range from interested, ecologically-oriented lay people to those whose economic support is directly affected, especially important are the planners and government administrators who have responsibilities in such matters. They must face the fact that, whether their concern is intra- or interregional or intra- or international, a fundamental land/population ratio must be maintained if population needs are to be met. Somewhere there must be adequate land space to produce the crops, raise the animals, grow the trees, and otherwise serve a given-sized population. Increasingly, the need to provide recreation space is emerging as a problem also, both locally with city parks and town forests, and nationally with a variety of types of recreation areas. The problem is exceedingly complex. Estimates of space needs become obsolete while the calculations are being made because capabilities and numbers are constantly changing. Yet we are becoming aware that we cannot go on indefinitely paving over the best agricultural lands, clearing more forest areas than we plant, or "developing" various kinds of sites with no regard for how they may change their surroundings. Moreover, the cost of resolving problem situations in our urban areas which result from unplanned or inadequately regulated growth is reaching proportions which indicate that restricting planning budgets is false economy.

The administrator and the planner must have comprehensive and current information if they are to have any hope of promoting orderly use of the land. There are many kinds of information useful to them and remote sensor data is only one. Planners have made use of one form of remote sensor data, airphotos, from the beginning, and periodic rephotographing of urban areas is a standard practice. But ever more sophisticated data is needed as urban problems become increasingly complicated. The rate of change in many localities has increased to explosive proportions, and it is very clear that the types and quantities of data which were used to come to grips with yesterday's problems simply will not suffice for today's. Developments in remote sensing offer the prospect of markedly greater variety and amounts of data. However, the task of learning how to use this data has only begun.

Land Use, General

A prerequisite to monitoring change in overall use of the land or developing a plan for improved land use is knowledge of the current pattern. We presently have maps of land use in this country, of course, and have had such maps for some time. The older the maps, however, the more unsatisfactory they are, in ways not apparent to the casual viewer. If the cartographer has done his job well, a map has a professional appearance and the originating agency name provides official status. It is when we look at the data used by the cartographer that shortcomings begin to appear. Methods of data collection*, the nature of the data (including its date), and sample size are examples of factors which often restrict the map's value for many would-be users. Yet it may truly represent the best obtainable at that time.

Consider the most recent government map of land use in the United States, which appears in the impressive *National Atlas of the United States* published in 1970. Compilation of the data and the actual cartography and publication took two years, so that the map is based on the best estimate obtainable of what the situation was two years before its publication. But this estimate includes some input from surveys carried out only at widely spaced intervals; some of the data used was at least 10 years old by the time of the map's publication. There is no intent here to demean this map; it is the best that we have ever had. The point is that the methods and types of data used will not suffice for future maps. Our needs have become more specific and immediate for most tasks.

Assessment and monitoring of land use on a national scale is likely to be carried out with different goals in mind than in regional or local cases. For most long-range planning purposes an up-to-the-minute national map is not necessary, if indeed even possible. There are large areas in which the use of land changes very slowly or not at all. In some areas, however, change on a local or even regional basis is occurring with unprecedented speed. In such areas especially, the need for "a better way" is evident in the form of distressing ecological problems that appear to be at best the product of lack of foresight.

Estimation of the environmental impact that a drastic change

---

*The common procedures followed by many nations to obtain land use information include mailed questionnaires, census interviews, and field mapping of sample areas.

of land use might have requires first an accurate knowledge of what is there. Operating on the premise that watching change occur would be beneficial also, a three-year project called CARETS was begun in an area around Chesapeake Bay designated the Central Atlantic Regional Ecological Test Site. As a jointly sponsored NASA/US Geological Survey effort, the project is testing the hypothesis that satellite-generated (ERTS) data could be useful both for an inventory of the land resource and for monitoring change in it (Alexander, 1972). Aircraft overflight data and ground data are being correlated with satellite data to enhance the utility of the latter as well as to provide ground "truth". Despite its lower resolution, the satellite imagery promises great utility because it is available promptly and will provide continuous coverage throughout the three-year period. Those of us who have watched industrial parks, residential areas, shopping centers, and freeways spring up in less than three years will realize how such changes alter the appearance of an urban area and its surroundings in that time.

Total regional land use management has to contend with a basic problem: the increasing competition for space. The inexorable spread of urban development, the possible need to expand agricultural acreage as the population increases, the dwindling wildlife habitat, and the desire by more people to experience more outdoor recreation already are in conflict. And it is not simply a task of trying to allot amounts of space or decide priorities; some land is not adaptable for agriculture and some could not be realistically assigned to wildlife habitat or outdoor recreation. Ultimately the use of the land in many parts of the country will change whether we make an effort to order the change or not. If we could make a beginning by achieving a better assessment of present use and capability, the possibilities for more ordered use would be much improved. Recent legislation such as the National Environmental Policy Act and similar state legislation is a move in this direction.

A number of studies have shown that land use mapping for various purposes and at widely varying degrees of detail is possible through remote sensing techniques. Assessing overall land use on small-scale imagery first became popular with the advent of astronaut photography and, more recently, ERTS imagery. One of the earlier efforts employed Gemini and Apollo photography of the Southwest from Southern California to Texas (Thrower and Senger, 1970). Several studies have sought to assess the potential of ERTS imagery. One used a mosaic made from many airphotos (fig. 4-1) which were assembled and then reduced photographically (Rudd,

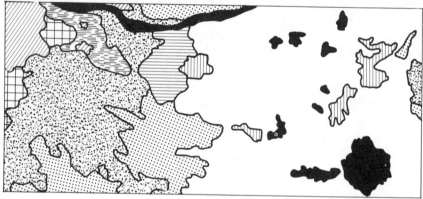

Legend
                                                                              Code
Forest   Continuous-Well developed 1.11

                Continuous-Poorly developed-Cutover 1.121

                Continuous-Poorly developed-Burned over 1.122

                Non-continuous-Forest dominant w/agriculture . . .   1.21

                Non-continuous-Forest and cleared/non-agri. 1.22

Agricultural   Cultivated crop and cultivated pasture 2.11

                Cultivated-Dominantly-But w/Woods 2.12

                Cleared-Unimproved pasture, etc. 2.2

Urban   Continuous-Concentrated 3.1

                Urban fringe 3.2

Non-Productive Water surface, mtn summit, poorly drained,
                riverine wooded tracts, and swampy bars 4.0

Figure 4-1: Mapping of surface features on a small-scale air photomosaic. Although the photo exhibits considerable complexity in some sections, distinctive tone-texture-pattern signatures make it possible to segregate at once some of the types mapped. Large cut-over areas and mountain summit are examples. (Courtesy of the American Society of Photogrammetry)

1971). Another photographically degraded some high altitude air-photos to the presumed quality ERTS would have, and then mapped on the degraded imagery (Raytheon, 1970). Studies in Chile (MacPhail, 1971) established the concept of *photomorphic regions* as entities being identifiable on photomosaics and having utility in land use mapping; areas in crops, in forest, or occupied by cities present contrasting signatures even though individual elements may not be discernible at the scale of the mosaic. This concept was used successfully in test mapping part of the TVA region (plate 8) from Apollo photos (Peplies and Wilson, 1970). For mapping of gross use categories, the limitations of resolution due to scale can be an asset in eliminating unnecessary detail.

Various problems were encountered in each of these studies, not the least of which was finding a satisfactory land use classification. Each of the efforts reports substantial success, nonetheless, in mapping from imagery with low resolution. For a first look, as in underdeveloped areas, or for the "big picture", then, satellite imagery offers significant promise. Chapter 5 will note a variety of successful applications of ERTS-1 imagery which bear out this promise.

The imagery sources for the examples of agricultural and urban land use studies which we will discuss below range from low-altitude aircraft to satellites. The scales of the imagery vary accordingly. At the present state of the art, the regional planner is likely to find satellite imagery valuable for more purposes than the urban planner, since the former's area of interest is larger and detail is less important. Both, however, will employ remote sensing most effectively if they employ a variety of scales, each best suited to the particular aspect of the task at hand. The advantage of timeliness which satellite imagery offers, though, will cause users to demand improvement of satellite resolution capabilities. The unexpectedly high quality of the imagery from ERTS which we now have in hand has made believers of many who questioned the whole idea. Having had a taste of sequential, potentially real-time imagery depicting over 10,000 square miles on one frame far more clearly than was expected, users such as area resource managers and regional planners will insist upon more and better satellite imagery.

Agriculture

Agriculture of some variety occupies on the order of ten billion acres of land over the world today. About 10 percent of the earth's land surface is cultivated, and these areas probably represent the most desirable, accessible, or readily usable land in their respective parts of the world. There are other areas that are believed to have possibilities—perhaps twice as much land—and some of them will no doubt be put to use as the pressure to feed the world's people increases. But most of these other areas pose problems of one sort or another. Some of these problems are human in origin, but the land itself must be amenable to cultivation, and most land is not. Perhaps the most obvious example of an agricultural land problem is climate. Crop agriculture is still predominantly at the mercy of atmospheric conditions, although in some parts of the world we have achieved limited success in diminishing the atmosphere's effects. The shape of the surface and the nature of the soil are additional limiting factors. Again, we have found ways to modify each of these where minor modification will suffice, but the majority of the land area of the world has problems of slope, climate, soil character, or drainage which make it presently unfit for cultivation. Even the underdeveloped areas with limited potential are not encouraging. Or so we believe.

The point has been made that our knowledge of the natural environment leaves something to be desired; perhaps there is more land suitable for cultivation than we think. Although it is unrealistic to imagine there may be great areas with excellent agricultural potential that have somehow escaped our notice, it is less so to suppose there may be many smaller areas which collectively would make a significant addition. Whether additional agriculturally potential land is "found" or not, however, a better knowledge of the additional 20 percent or so believed to offer possibilities is a worthy goal. If only half of it were ultimately to be put to use, we would have doubled the world's land in cultivation, even though much of this land would have limited productive capacity.

There are many reasons why the marginally cultivable lands of the world are not in use, and many lands would continue to lie idle after the most exhaustive survey. But inadequate knowledge is an important deterrent in many cases. Perhaps remote sensing, by reducing survey costs, by making surveys possible where they have been difficult, or by revealing characteristics that conventional surveys missed, may make a major contribution to agricultural development. For nations in which a massive application of agricultural technology is impractical, a better knowledge of the agricultural resource base could be extremely valuable.

Plate 1: A color IR photo of part of Denver, Colorado. Lawns, grassed areas, and trees are shown in red. Brightness of red tone indicates differences in plant vigor and extent of cover by live, healthy vegetation.

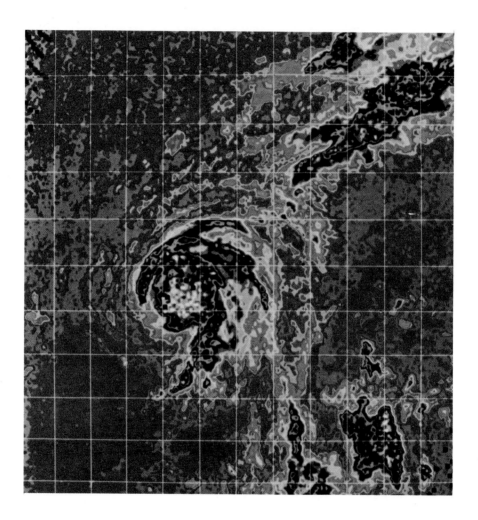

Plate 2: Hurricane Camille, 1969, as viewed by Nimbus 3. This thermal IR image has been modified for color display to emphasize cloud temperature variations which indicate differences of storm activity. (NASA photo)

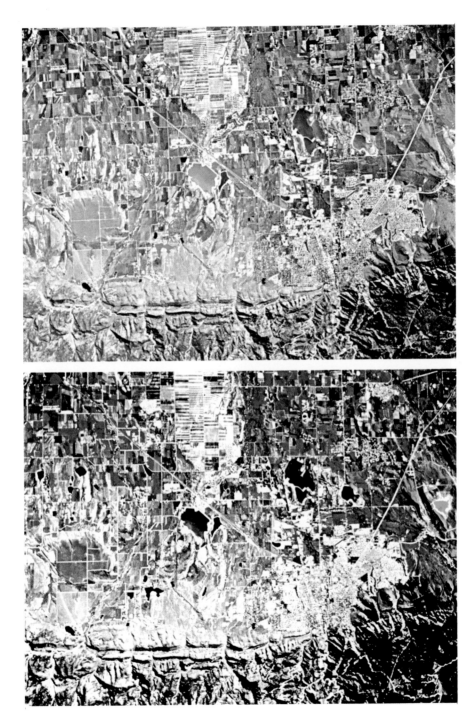

Plate 3: Conventional color and color IR view of Boulder, Colorado. The contrast requires a second look to insure that the one is a color photo at all. Note the contrast in sharpness also; the photos were taken at the same time. (NASA photo)

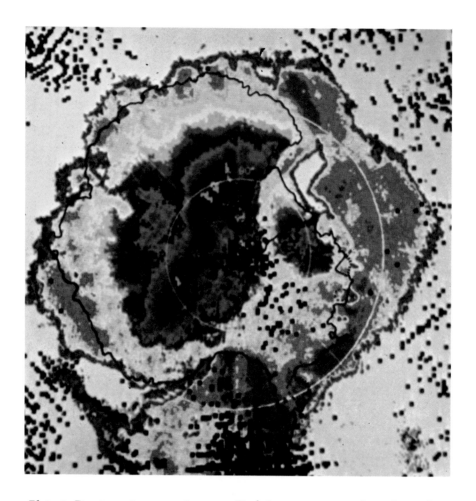

Plate 4: Passive microwave imagery. Both images were made with an Electrically Scanning Microwave Radiometer (ESMR) from NASA's Nimbus 5 weather satellite in circumpolar orbit. Classes of brightness temperature may each be assigned a color and variations in radiation then can be depicted in map form. The two views of Antarctica illustrate changes in the

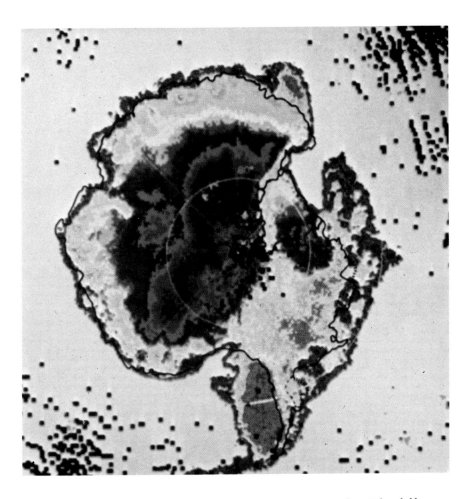

amount of ice as "summer" comes to the continent (right). The different colors indicate, in part, differences in the nature of the surface material. Black spots represent data gaps. (Courtesy of Aerojet Electro Systems and NASA)

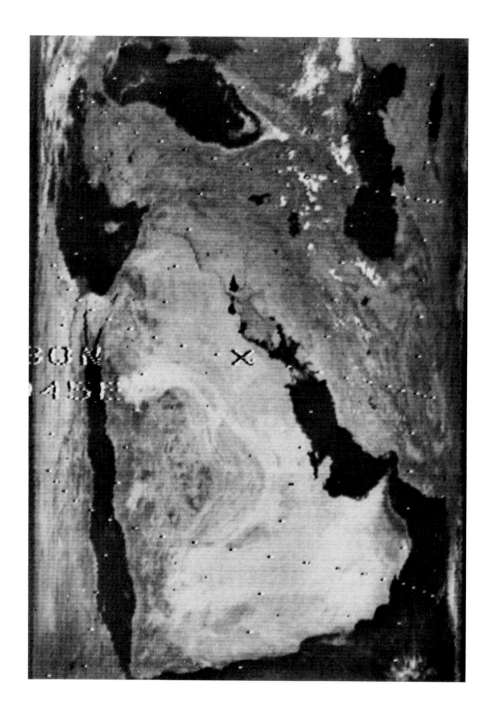

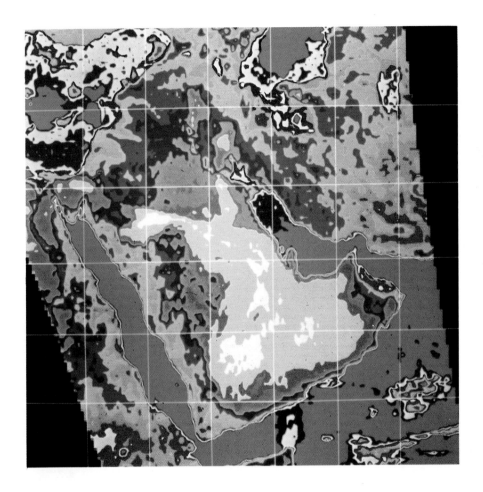

Plate 5: Nimbus 3 view of Saudi Arabia. Gross rock types and surface materials are among the factors which cause the patterns on this near IR image. Color enhancement (right) points up these patterns. (NASA photo)

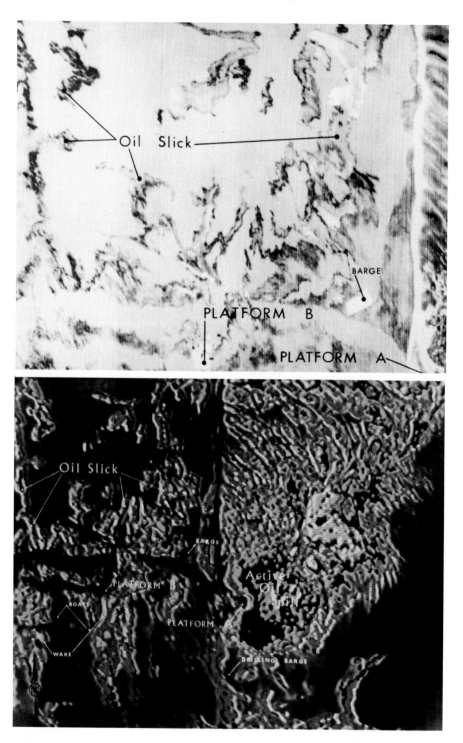

Plate 6: Color enhancement of a thermal image of an oil spill. The black and white thermal scanner image is more revealing than a conventional photo, but color enhancement of the image provides a further refinement. (Courtesy of John Estes and Berl Golomb)

Plate 7: The ATS III view of earth from 22,300 miles out. Since ATS III is positioned over the equator, South America (lower right) is better shown than North America (upper left). (NASA photo)

Plate 8: Evidence of variable land use on space photography. Varying degrees and types of activity cause signature contrasts which form the basis for bounding regions on this Apollo 9 photo of western Mississippi,

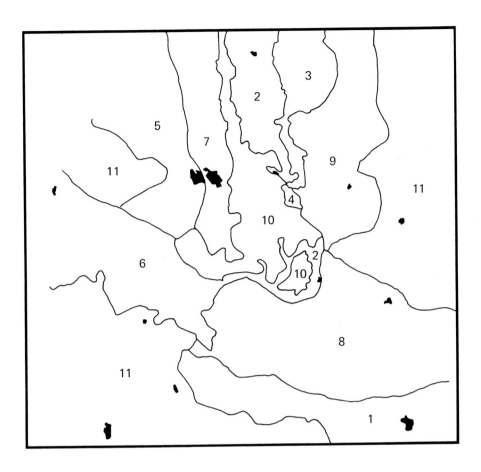

1	Black Belt Area
2	Moulton Valley Region
3	Bankhead Forest Area
4	Russellville Mining Section
5	Tennesee Valley Small Farm Region
6	Tennesee River Floodplain Area
7	Tennesee Valley Large Farm Region
8	Tombigbee-Tennesee Region
9	Upper Coastal Plain Farming Region
10	Little Mountain Woodland Area
11	Undefined
⬛	Urban

northern Alabama and central Tennessee. (Courtesy of Robert Peplies/NASA)

Plate 9: A color IR photo illustrating crop distress. In the square field near the photo center below the diagonal road several areas of corn blight can be seen on the color IR photo. They are much less evident (if at all) on the conventional color photo above. (Courtesy of Laboratory for Applications of Remote Sensing, Purdue University)

Plate 10: Tree distress indicated by color IR. Several of these ponderosa pines have been affected by a needle blight; the lighter the shade of pink, the more severe the stress. Healthy trees are red. (Courtesy of U.S. Department of Agriculture, Forest Service)

Plate 11: Urban neighborhood variability (color IR photo). Even at this scale, differences in lot size and surroundings between residential sections in the lower right and those in the lower left or upper right are evident. (Department of Geography, University of California, Riverside, and the Geography Branch, Office of Naval Research)

Plate 12: A "created" color IR image from ERTS-1 data. The southeast corner of the Imperial Valley of California is lower left; the Colorado River separating California and Arizona is in the middle of the photo. Cropped fields stand out even more sharply in color IR. (NASA photo)

Plate 13: The San Francisco Bay area as viewed from Skylab. This color IR photo of west central California was made with one of the cameras in the 6-camera system. Clouds reach in from the sea to the Golden Gate, but the urban areas of San Francisco, San Jose, Oakland, etc., are clear as is the northwest portion of the San Joaquin Valley east of the Coast Ranges. (NASA photo)

The Harvest

Over the years there have been those who, after assessing the then-current agricultural production potential, have predicted future disastrous food shortages on a national or world basis. From time to time there actually have been such tragic developments, accompanied by great suffering, and occasionally they have occurred at the time they were predicted. But one factor which has repeatedly confounded the prophets, especially those who have looked ahead as much as a decade or a generation, is the way production per acre has increased. Often new varieties of crops, improved cultivation methods, or advances in equipment have resulted in far greater production on essentially the same amount of land*, and predictions have thus been rendered totally inaccurate. Perhaps this fortuitous trend will continue and the forthcoming populations will somehow be fed with varying degrees of adequacy as in the past. Assessments which indicate that our production gains are being consumed as rapidly as they occur and the unprecedented dimensions of the population increase we face, however, cause uneasiness. It has been suggested that what is going to be required in the next 30 years is no less than a doubling of the production capability that we have developed since agriculture began (Chapter 1, National Academy of Science, 1970), despite current population trends in the U.S. There is great potential for agricultural improvement in the underdeveloped nations of the world. Since these areas include over half the world's people, substantial change would greatly modify their situation. But increased productivity will be necessary even in the most technologically advanced areas, too.

Harvest size could be increased significantly by more thorough monitoring of crop vigor and health. Color infrared pictures taken from two miles above a field may reveal crop stress before the farmer can detect it while walking through the field (plate 9). With some crops, disease shows up first in the ultraviolet. Remote sensing should be specially valuable in situations of extensive agriculture where fields are so large that surface monitoring is difficult. Remote sensing has utility also, however, for intensive

*According to one study, U.S. farmers have been producing increasing amounts of food and fiber for some years from farms which collectively occupy less and less acreage. See Hart, J. F., "A Map of the Agricultural Implosion", *Proceedings of the Association of American Geographers* 2 (1970): 68−71.

agriculture where even a small area loss means a large loss quantitatively. Whether the problem is insects, disease, or simply lack of sufficient water, its early detection can reduce the loss and thus increase the harvest. Sections of a field which show up as problem areas repeatedly can be bounded precisely and subjected to analysis and remedial treatment.

Residual soil moisture can be of significant importance to potential yield. Several sensors (scanners, radiometers) have demonstrated their ability to detect soil moisture variation; and if signature parameters which indicate satisfactory moisture conditions can be established, reduced yields or outright loss due to moisture shortages should be preventable.

Improvement of the harvest estimate is a never-ending task; the best we have been able to do continues to prove unsatisfactory for some needs. Yet estimates are essential for many reasons, and advance preparation of the massive systems and facilities for storage, processing and distribution is only one example. The problem of oversupply of some crops and undersupply of others can be alleviated with a better information base since planting times differ in the several parts of the nation—and the world. A rather specialized example of supply estimation is provided by the raisin grape harvest (Chapter 4, National Academy of Science, 1970). A proportion of the grapes in California are picked early for raisins and laid out in trays between the vine rows to dry. The remainder are left on the vines to be picked later for other purposes. In the past a combination of factors such as sample ground counts and past performance figures have been the basis for decision by the grower when to cut off the raisin grape harvest. Too often, over- or undersupply has been the result. A growers' association decision to spend $50,000 on an aerial survey, which provided photos on which the trays lying in the fields could be counted, is estimated to have saved $5,000,000 in the more effective disposition of the crop.

It would be impossible in this short book to explore the theoretical economic reasons why more accurate and complete knowledge of the acreage of major crops, as they approach harvest date, might contribute to greater price stability. Or how crop vigor data might improve harvest estimates and thus provide some foreknowledge of what prices could be expected at harvest—and so on. But intuitively it seems that such information *could* produce economic benefits if used correctly. More important, foreknowledge of the way the world's crops are progressing and the likelihood of their adequacy would surely be beneficial in the alleviation of impending shortages. This kind of information would be extremely valuable, even vital, in helping to reduce the dimensions of anticipated future world food problems.

Is it realistic to expect such information from remote sensing? We are a long, long way from anything even approaching accurate, timely forecasts of world harvests. Presently the cost of using current aircraft coverage of the nation's (let alone the world's) total agricultural areas, especially on a repetitive basis during the growing season, is prohibitive. Satellite sensors are not sophisticated enough to provide operationally the scales and types of imagery that we would need, although they can provide repetitive imagery of the whole world from one platform at reduced cost. The identification of individual crops on remotely sensed imagery is a difficult task; definitive signatures for most crops have yet to be established. We are still in the process of learning to use this new tool. And the task of analyzing the volume of data involved in coverage of this nation's agricultural areas alone is beyond the capability of the presently available trained labor force. There are other limiting considerations, but those mentioned indicate the nature of the problem.

The point has been made repeatedly that neither aircraft nor spacecraft imagery alone is likely to be as effective for most substantial tasks as joint use of the two. The prohibitive cost of the former and the state of the art of the latter will become lesser obstacles as their most effective joint uses are identified and the sophistication of spaceborne sensors increases. The cost factor is one that requires special consideration, however, and is discussed in Chapter 6.

Personnel and Automation

The number of people presently employing remote sensing on an *operational* basis is relatively small. Some military and National Weather Service personnel constitute the most easily identified groups. The majority of the others involved with remote sensing carry on their activities on an experimental or research basis. Should government agencies decide to employ remote sensing more widely, they will have to educate the necessary personnel. Relatively few of the present federal or state government employees are versed in the technique and its applications. Universities are incorporating instruction in remote sensing application to the various disciplines so that some future graduates in fields such as forestry, agriculture, and geography will be familiar with the concept. The number of students is not large, though; and, with the quantities of data being returned from ERTS-1 and Skylab and forthcoming from future satellites a major effort at implementation could be frustrated by a shortage of trained manpower.

There is not likely, however, to be a concerted effort to produce large numbers of specialists in remote sensing in any way commensurate with anticipated data quantities. Early in the mushrooming of this field it became evident that machines would have to be enlisted to make effective use of the quantities of data in prospect, and research on automated analysis of data has been underway for some time. Although most earth resources data is not as perishable as weather data, some is useful for only a limited period; and, as with weather satellite data, sheer volume will limit its usefulness without computer help. To appreciate the volume of data we should remember that it is available in two formats: in collective form as imagery and as individual bits stored on tape. It is easy to forget (such is the photographic quality) that each of the ERTS imagery examples in this volume was made from millions of signals received by the sensor in the satellite and then transmitted electronically to ground receivers before they became a picture. One ERTS frame (image) is made from over seven million data bits, transmitted to earth and stored on tape and then played back onto electronic equipment to create the image. For some purposes the imagery format is preferable, but for others it is better to use the data on tape. The combination of the number of images potentially available plus the unimaginable quantities of data in bit form is what necessitates machine assistance.

Two concepts discussed earlier are being utilized in the effort to apply remote sensing techniques and automation to agriculture studies: signature identification and multiband spectral analysis. Crop identification on airphotos has always been plagued with problems. For crops of similar photo appearance, success has typically required assistance from other information sources such as ground sampling, past history, or correlating photo dates with knowledge of crop seasonal development. These methods will continue to be useful, but different procedures, including automation, are needed for using remote sensing data most effectively. Tonal differences on a photo associated with different crops can be detected by photoelectric devices and converted electronically into varying electrical signals, storable on tape. Or the taped data direct from the sensor may be used. A computer system can then be programmed to read the tape and print out a selected symbol—say, C for corn—for each similar stored signal. With the system designed to include the photo address of each signal as it is stored, the replay printout takes on the format of a map (fig. 4-2).

Some devices such as a microdensitometer can detect tonal differences on a photo transparency more subtle than the eye can see, and provide some advantage over a human analyst. But crop signatures based on tonal differences on conventional film remain elusive, so the combination of conventional and infrared film in signature search has been tried. Corn might show up on both types of film as light in tone, wheat might be light on one and dark on the other, oats dark on both. With only one film type, only one of the three is identifiable, but with two film types all three are distinguishable. In this oversimplified example much is left unsaid, but the basic principle of multiband analysis is illustrated.

In the real world, far more than three crops are possibilities; and, although texture, patterns and many other parameters in addition to tone are used as signature elements, the problem remains a difficult one. We have tried truly "multiple" band analysis by comparing photos made with only green light, or blue, or red, with one another and with ultraviolet, infrared, microwave, and radar imagery in seeking combinations that will yield distinctive signatures.

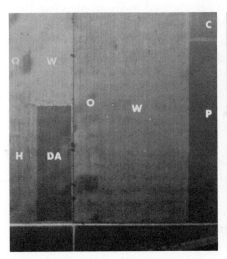 

Figure 4-2: A computer printout of crop identification. The letters on the photo identify the crops; O for oats, C for corn, W for wheat, etc. The computer printout sheet shows how the automated system interpreted the remote sensor data. (Courtesy of Laboratory for Applications of Remote Sensing, Purdue University)

More encouraging, however, is automated signature identification that does not use the data in an imagery format. One of the major efforts of this type is being carried on at the Laboratory for Application of Remote Sensing at Purdue University.

A multispectral scanner is capable of collecting radiation from an agricultural scene in a number of bands simultaneously. Ultraviolet, blue light, green light, red light, and infrared radiation can be received and stored separately. The signal response of a given crop in each of these bands can then be examined for signature identification but in a different manner. The computer can be programmed to examine the signal electronically, bypassing the conversion of the data to imagery form. Moreover, the signal can be subjected to a very sophisticated scrutiny by having numerous comparative queries built into the program, queries which investigate in *which part* of the green band the signal is strongest, for example—an intraband analysis to go along with the multiband analysis. And with computer help it is done in far less time than you have here taken to read about it.

The successful wedding of automation to remote sensing and further developments in each may produce advances in agriculture unimagined a short time ago. Aside from the improved inventories of the nation's agriculture, the possibilities for improved management are especially exciting. Some far-sighted individuals visualize the monitoring of irrigation practices, both to assure adequate soil moisture and, in the arid lands, to avoid excessive watering of soils subject to salinization. Fields could be periodically checked for need of fertilizer applications. Not only could insect or disease problems be identified, but monitoring the effects of control measures could reduce wasteful application of chemicals. In some cases, the writeups read like science fiction (Park et al., 1968), but science fiction sometimes has a way of becoming reality. Perhaps some ideas will never see operational application; however, they are likely to be replaced by others that now would seem just as unrealistic.

Forest and Range Management

Extensive areas of the forest and rangeland of the world are subjected to little or no management. The vast forests and savannas of the tropics are situated in nations that cannot provide the kind of monetary outlay required for an effective management program. Fires, disease, insects, and misuse problems occur periodically and assume disastrous proportions, often because they reach an advanced stage before the existence of the problem is realized. In the early 1960s, a southern pine beetle epidemic in Honduras affected over four million acres of that nation's forests, killing trees which represented almost nine billion board feet of pine (National Academy of Science, 1970). There are great forests in the high latitudes also, the majority of which are subjected to only modest management effort. The sheer size of these forests, which stretch completely across the North American and Eurasian continents, poses management problems which are a challenge for even the better endowed nations within whose borders they lie. To date we have made limited use of both tropical and high latitude forests, a situation not likely to change radically in the near future. Access is only one of the problems which make them less desirable than mid-latitude forests. Still, they constitute a majority of the world's forests and represent a considerable resource potential. They and the tropical grasslands are also major wildlife habitats.

In the mid-latitudes more and more nations are adopting the attitude that forests and rangeland also produce a crop and must be managed. An element of stewardship has assumed some importance in their management, but pragmatic considerations have been the basis for most of the effective action.

In forestry, such questions as what species are present, how much timber is there, and what are the growth rates have long been answered in some nations through field studies complemented by aerial photography. Planning the logging operations, preparation for fire control, and the assessment of the vigor and general health of the stand are additional considerations. Advances in remote sensing promise assistance on all of these matters. Multiband analysis for species identification, thermal scanners for fire detection and assistance in fire fighting, radar's all-weather capability for obtaining coverage of areas which resist photographic efforts, and photographic infrared indicators of vigor are examples that have been mentioned. Since insect or disease symptoms commonly first appear in tree crowns, and early detection is critical to the success of remedial action, the importance of reconnaissance from above is underscored (plate 10). In the case of some diseases such as root rot, the pattern and size of the affected areas is such that even satellite imagery is believed to have potential, especially for remote areas.

We commonly think of forest and range as separate categories although we know that there are extensive mixed areas which might be called either. In fact, grazing is often cited as an example of multiple use in the national forests of the United States. The forests and rangelands constitute the habitat of much of the wildlife, but they also are used by domesticated animals. The term "range" as presently used in this country is very inclusive. Basically it refers to nonarable land which is capable of producing wild vegetation usable as food by animals. We might note, in passing, the implication that even if the present vegetation is inadequate or undesirable, it may still be classed as range if it has the potential for more suitable vegetative cover. More to the point, however, is that not only the semi-arid grass and brush lands that most of us visualize, but also open forests, mountain meadows, and tundra areas are included in this term. Nearly half of the land surface of the globe can be called range, given such a broad definition.

Both for the health of the animal life and the maintenance of the quality of the grazing areas, care must be exercised in the use of range lands. Carrying capacity is an example of the many considerations to be taken into account. Aside from determining the number of animals that will use an area, knowledge of the amount, nature, and condition of the vegetative cover is basic. The area will have variable capacity at different times of the year and in succeeding years. Changes in plant species, the percent cover, and the soil moisture content are variables which, if not monitored, can result in unintentional misuse. Research which relates soil and foliage temperatures to soil moisture content and transpiration rates is an example of how remotely sensed data may be useful in the prediction of range capacity. Greater moisture availability should make for higher percent cover and thus greater biomass. Maintaining a reasonable ratio between the number of animals and the capacity of the range vegetation they feed upon should be made easier with the improved data base remote sensing can help provide.

There are other considerations, however, which transcend the maintenance of rangelands for continued support of grazing animals. The term "range resources" can include animal products, wildlife (including fish), timber, soils, and even minerals. Such a variety of resources increases the potential returns of careful management. Proper care of the rangelands can also exert a positive favorable influence on adjoining areas having other uses; negligence or outright misuse of them will result in equally profound undesirable effects. The amount of land involved underscores the need for greater attention to and care for the range resources of the world than we have accorded them in the past.

Urban Land Use

Few would seriously challenge the statement that one of our most pressing problems today is improving the quality of life in urban areas. The problem itself is not really so recent; students of cities and their development tell us that crises have occurred throughout urban history. However, there are components of today's urban problems that were not present in earlier crises. Not only are the numbers of urban dwellers and the size of urban areas unprecedented, but also urban residents are better educated and more vocal than ever before. They are demanding to know why, in this day and age of miracles routinely performed by science and technology, their problems are not being solved with dispatch.

Several decades ago the entry "city planner" began to appear on the personnel rosters of city governments. Although many of the problems which plague our cities today were not foreseen, at least in detail, it had become apparent that cities could not be allowed to grow further without a better effort to plan that growth. In brief, the planner's task was to provide a program for guiding growth and development. Eventually he was made responsible for the improvement of the urban environment. Most cities of any size have a planner today; the larger ones have substantial planning staffs.

The Nature of the Problem

What are the considerations for guiding development and improving the environment of urban places? Some, such as adequate transportation facilities, are self-evident. Millions of people must be able to move daily from home to work and back, goods and people must come into or leave the city, intracity movement of all sorts must be accommodated—all in reasonably expeditious fashion. A suitable transportation network must be provided, maintained, and continually modified to offset change.

Another very basic consideration is simply how many people are where. This seemingly easy question is in fact difficult to answer because of the increasing mobility of urban dwellers. Yet solving problems like providing adequate police and fire protection or modifying utility facilities to fit changing demand depends upon the answer. The problem of zoning requires, among other things, an answer to another apparently simple question: What is the existing land use in the area? Also, approving plans for new subdivisions requires considering where, for example, to locate new storm sewer systems, and how they will affect existing natural and artificial drainage systems. Recently the desire to focus efforts on improving housing quality has raised another question: How do we identify "blighted" areas and where are they? If we add, "How can planning lessen the effects of pollution?" and, "In which direction(s) is the city growing and is this desirable?" we have a sample of the questions planners are asked to consider.

To be sure, these and similar problems are the responsibility of many people. The planner's contribution is only one, but planning requires, above all, looking ahead. It is not enough to know the current situation; he must forsee both the possible range of alternative futures and what is most likely to happen. However, obtaining accurate information to help solve any of the above problems means delays—in some cases a year or more for a complete view. In most cities the nature of the current data base and the methods used to obtain it are inadequate to provide an accurate up-to-date view of the situation. It follows that they are at least equally inadequate for the task of planning. This deficiency is only one of the reasons that today's urban problems are not being solved quickly enough, but it is a very basic one. And it is here that remote sensing may make a contribution.

Applications Research

A knowledge of what classes of housing exist in a city and of their spatial distribution has long been useful to city governments in assessing property value for taxation. In recent years the federal government has asked cities to identify poverty-stricken neighborhoods for the purpose of allocating financial assistance for their improvement. Some cities watch for indications of blight with the intention of seeking timely remedial solutions. Yet, current city maps accurately depicting classes of housing are surprisingly rare. Conventional aerial photography does not depict such features very effectively; field enumeration surveys are costly, subjective, and very slow; and tax records or like documents may be unreliable.

In view of this situation, several efforts to define housing classes and neighborhood quality on remote sensor imagery have been made (plate 11). Mapping on color infrared photography has been more effective than on conventional, partly because features are more sharply defined and the presence of vegetation is highlighted. Studies using multiband analysis also have proved encouraging. In these and other similar studies, house size and shape was often less important than such surrogates (substitute indicators) as street characteristics, yard size and landscaping, presence of refuse and untended areas, and various hazards. Results from studies based on thermal IR and radar imagery were more of a qualified success, since resolution on neither imagery form was comparable to photography. Subsequent sophistication of these sensors may overcome this drawback, but their greatest value may be their utility in haze or after dark, especially for traffic studies. Radar's greater areal coverage from a given altitude may be an advantage at times, but it is a questionable trade-off for resolution.

Some studies have concentrated as much (or more) on methodology as on sensor systems. It has been shown in assessing neighborhood quality, for example, that using only a fraction of the conventional signature components can produce acceptably accurate results if those components are judiciously selected (Horton and Marble, 1969). Since these components are visible on imagery, remote sensing could make a contribution here.

Numerous efforts to map urban land use have been carried out, employing a variety of imagery forms and scales of imagery. Greater success has been achieved with color IR than with other imagery forms; the superior resolution of photography continues to be basic. Scale is harder to summarize. Some early success in using orbital imagery was reported, but commonly the purpose was to monitor gross change or to view the urban unit or units in a regional framework. With the more recent ERTS data, some success in more detailed analysis has been achieved by electronic enlargement of the image. Although it is not reasonable at this time to expect space-generated data to replace large-scale data for detailed mapping, indications are that it is a valuable supplement. A more fundamental problem is that the compatibility of remote sensor data and other data from more routine sources has not been ideal. Efforts to incorporate the two into improved information systems have been only moderately successful. A basic reason is the state of the art of the information systems themselves: models which make the most effective use of remotely sensed data are as yet unproven or have yet to be designed.

The Potential

Most students of urban planning who are informed about remote sensing feel the technique promises to make a greater contribution to their field. A basic need is timely data in quantity and variety, which is a demonstrated capability of remote sensing (Moore & Wellar, 1969). Examples where timeliness is critical include traffic studies, especially those where unforeseen or new problems emerge, some varieties of pollution problems, the emergence of hazards or the occurrence of disasters. Even simply monitoring change requires timely data, because of our increased ability to change the landscape rapidly. Other cases in point are the all-weather capability of radar, the hitherto little-used night view of situations which both thermal IR and radar afford, and the capability of imagery in general to "freeze" the traffic or parking situation at selected times.

Data gathered by remote sensing has an objectivity which is advantageous. Analysis of the data can be rechecked without resurveying, and a common data base has been used overall. Moreover, there are kinds of data not presently used, and therefore not collected, which will become desirable in future years. Sensors collect data indiscriminately rather than selectively—and make a permanent record of it.

At present, we make limited use of advances in remote sensing in urban studies, partly because many such advances do not have demonstrated operational capability and are costly. In some cases, there is also simply the reluctance to become involved with a new concept. Greater use of this tool seems inevitable, but it will depend upon further research in utilizing the data, improvements in sensor capabilities, the development of information systems which can implement the data, and the establishment of sound cost-benefit comparisons. Apparent costs often are deceiving. Examples abound of the routine use of systems and techniques once considered too expensive that actually have proven to be economical.

Chapter   5

The Development
of Remote Sensing
From Space

During the early 1960s, a few people who were aware of the developments occurring in remote sensing began considering how they might be applied to earth resources studies. The growing successes with satellites and the possibilities hinted at by evolving sensor systems implied a potential for study of the earth far beyond anything ever before attempted. By the mid-1960s, although there was almost no effective utilization of satellite-generated imagery for study of the earth and few scientists were familiar with remote sensing, both national and global circumstances indicated a desirability for moving ahead.

In 1966 the Earth Resources Observation Systems (EROS) program was established under the U.S. Geological Survey in the Department of Interior. It was but one of several early attempts to employ the potential capabilities of remote sensing. In less than a decade, we have made astonishing progress to reach the present state of the art.

It was immediately apparent that the effective application of remote sensing required that many people be educated. Most of those who understood the technique itself were not versed in substantive fields such as the natural sciences. At the same time, few geographers, geologists, foresters, or biologists knew anything about remote sensing. By various devices such as symposia for idea exchange, institutes and short courses, more available research, money, and cooperative efforts between NASA, other involved federal agencies, and the universities, the problem was attacked. The success of the effort may be measured in various ways. For example, by the time the first earth resources satellite, ERTS-A, was ready for launch there were so many proposals submitted for research projects to utilize the data that only a fraction could be supported.

The earth's surface now has been imaged from space for well over a decade, at first incidentally but lately more and more deliberately. The first weather satellites returned, along with the views of cloud patterns they were designed to collect, a fuzzy, indistinct view of the surface. As sensor systems on weather satellites became more sophisticated, depiction of the surface improved along with that of cloud patterns. Some surface information, in fact, became the object of sensors added to later weather satellites as land and water temperatures, snow and ice cover, and similar data were sought. We began to get photographs also, taken by the astronauts or by automatic camera systems from manned spacecraft, whose quality heightened interest in studying the earth from space.

Although there already have been many studies of the earth's surface and its resources based on space-generated imagery, the data collection effort that produced the imagery was not solely or even primarily for earth resources study. The first satellite equipped with a sensor package designed specifically for such study is the Earth Resources Technology Satellite (fig. 5-1). ERTS and the Earth Resources Experiment Package (EREP) of Skylab, show what useful information about the earth we can expect to obtain from orbiting satellites at the present state of the art.

ERTS

On July 23, 1972, ERTS-A, mounted on top of a McDonnell-Douglas Delta rocket, lifted from the launch pad at Vandenberg Air Force Base and, in so doing, became ERTS-1. ERTS-B, a planned sister satellite, will retain that designation until it is functioning in orbit and earns the designation ERTS-2.

Orbit

The orbit selected for ERTS-1 determines the satellite's period of revolution and the area covered by its sensors. At an

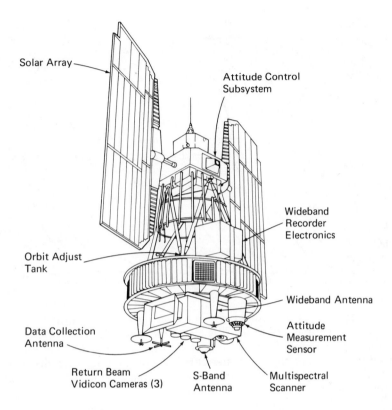

Solar Array

Attitude Control
Subsystem

Wideband
Recorder
Electronics

Orbit Adjust
Tank

Wideband Antenna

Attitude
Measurement
Sensor

Data Collection
Antenna

Return Beam
Vidicon Cameras (3)

S-Band
Antenna

Multispectral
Scanner

Figure 5-1: The ERTS-1 satellite. The common lineage of this satellite and
the Nimbus V (fig. 3-4) is evident. (NASA photo)

altitude of 492 nautical miles, ERTS-1 makes one revolution of the
earth each 103 minutes, thus completing about 14 orbits in 24 hours.
Since the earth is rotating beneath the satellite, and makes one full
rotation in 24 hours, it may seem that total earth coverage would be
achieved each day. At the altitude selected and with the
characteristics of the sensors onboard, however, there are large gaps
in coverage between successive orbit swaths. The ground width of
the swath covered by sensors is approximately 100 nautical miles,
and 14 swaths miss the over 21,000 nautical miles of the earth's cir-
cumference by a substantial margin. Actually, successive orbits are
about 26° apart at the equator, a far greater distance than 100
nautical miles. As the earth completes one rotation and the satellite
continues its revolution around it, the ground paths swept out by
the sensors are different from the previous ones and thus cover
some of the areas missed during the previous 24 hours (fig. 5-2).
This shift of coverage continues until the satellite is once again over
the first path and the cycle begins all over. A complete cycle requires

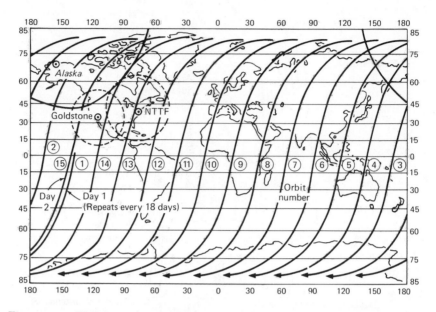

Figure 5-2: ERTS-1 orbit diagram. The lines represent the ground trace of the satellite's course overhead. The slight inclination of the trace near the equator reflects the 9° tilt from a north-south orbit plus the rotation of the earth during the orbit. The greater curvature of the trace near the poles is due to the map which exaggerates east-west distance at high latitude. (after NASA diagram)

18 days, and although during this period a given location will be imaged marginally a number of times, the satellite will be directly over it only once every 18 days.

Another aspect of the timing factor provides a hint of the intricacies of space science and the planning which went into this venture. Since the sensors onboard ERTS are dependent upon reflected solar energy for imaging, the satellite will have to be located over desired target areas at times of satisfactory illumination. The 48 contiguous United States, for example, extend from about 25°N. latitude to 49°N. latitude. On the southbound sweep ERTS's near-polar orbit crosses the 50°N. parallel at 10:00 a.m. local sun time* each pass, so that the United States is always being imaged at the same time of day. Since ERTS is programmed to cross 50°N. southbound at 10:00 consistently, other locations at the same latitude are passed over by the satellite at near 10:00 a.m. local sun time, also, as the earth's rotation brings them around beneath the

---

*Sun time is the time registered by an accurate sun dial. The standard time concept utilizes zones throughout which the accepted time is the sun time on a designated meridian.

satellite's orbit. At other latitudes the satellite is overhead either earlier or later; when southbound the satellite crosses the equator at 9:30 a.m. local sun time, for example. Such an orbital arrangement is called sun-synchronous. Since the time of day on earth is based on location with respect to basic earth-sun alignments, selecting a given satellite orbit position with respect to this alignment is in essence the selection of the time format for overflight of each location.

There is considerably more to a sun-synchronous orbit than is immediately apparent, however. One additional consideration will illustrate the point. The original orientation of the satellite orbit plane to the sun may achieve a desired time arrangement, but how can we keep it? The satellite orbit is established by creating necessary relationships between the velocity and mass of the satellite, its altitude, and the mass of the earth; the sun exerts essentially no influence on the satellite's orbit. Hence, as the earth revolves around the sun, the orbit plane will maintain the same relationship to the earth but not to the sun. The desired time relationship, a sun-dependent thing, would thus be lost. The solution to the problem is to select an orbit which will keep the satellite orbit plane orientation changing so that its attitude to the sun remains constant (fig. 5-3). ERTS's orbit is tilted about 9° from a true polar orbit in the retrograde (westward) direction. The effect is to cause the orbit plane to precess (move) eastward about 1° per day thus keeping the same orbital plane-sun alignment throughout the year.

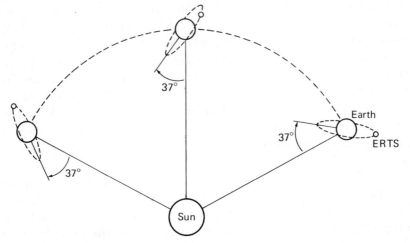

Figure 5-3: ERTS-1 orbit-plane orientation to the sun. The satellite orbit plane makes one complete rotation during one revolution of the earth around the sun. (after NASA diagram)

The cost of this arrangement is the inability to cover some 8° of latitude around the poles, so that ERTS does not provide total global coverage. The timing factor is important enough, though, that it is reckoned to be worth the cost. The illumination of a given scene must not only be adequate, but also as consistent as possible. Differences in the angle of incident solar radiation can alter the spectral response from a target, thus introducing a variable not due solely to change in the target. For example, an expanse of natural vegetation could produce varying signatures because of changes in illumination rather than changes in the vegetation. Seasonal changes in illumination cannot be eliminated, but limiting the effect of time-of-day differences on illumination can.

Sensor Platform

ERTS-1 is a modified Nimbus satellite, the type which has proved itself as a successful research weather satellite. It weighs about one ton and its most striking visual features are the two paddle-like solar array panels on opposite sides (see fig. 5-1). Solar energy is converted into electrical energy by these panels, providing a continuously renewable energy source (in conjunction with the satellite's eight nickel-cadmium battery modules) for the onboard equipment. The upper portion of the satellite also contains the attitude control subsystem, used when the spacecraft (and therefore the sensors) become disoriented with respect to the earth's surface. The sensor systems are clustered in the base of the vehicle, as is the equipment for transmission of the data to the ground. Various factors contributing to orbit decay may be corrected by a command from the ground which activates orbit adjust thrusters. To date they have been little used. The onboard fuel supply for these engines is finite in amount, though, and in time its exhaustion could end the satellite's usefulness. Failure of some of the other equipment has occurred, but the hoped-for life of the satellite at this writing is several years.

Transmissions from ERTS are received at three widely separated locations: Alaska, California, and Maryland. The distance between these stations plus the configuration of the ERTS-1 orbit originally allowed transmission of data on 11 of the 14 daily orbits; data collected on the other three orbits was stored onboard for subsequent transmission when the satellite was again in range. As discussed below, this capability has been reduced by malfunction.

Canada also has established a receiving station at Prince Albert in central Saskatchewan. All transmissions are forwarded to NASA's Goddard Space Flight Center at Greenbelt, Maryland, since the only ERTS-1 data processsing facility is located there. At Goddard the transmissions are processed into various formats, including edited and rectified tapes and film images. These are sent to the Department of Interior EROS facility at Sioux Falls, South Dakota, where they are made available to users.

Sensor Systems

ERTS-1 carries two remote sensory systems, a multispectral scanner and a three-unit television camera system. The two systems sense the same area, but they have complementary roles. The cameras image the total scene simultaneously by shuttering the image screen and therefore do not experience the blurring of parts of the scene which a slight pitch or roll of the spacecraft will cause on scanner imagery. The scanner requires a short time to image the whole scene, as the nonsquare image shape, reflecting the satellite's motion during the image-acquiring period, attests. The multispectral scanner has higher spectral fidelity, however, and the registry of the four images it collects simultaneously makes them easier to combine. Resolution is nearly comparable on the two systems.

The multispectral scanner onboard ERTS-1 is an optical mechanical scanner which senses simultaneously in four bands of the spectrum. Separate data are collected at wavelengths of .5 to .6μm (green), .6 to .7μm (red), .7 to .8μm, and .8 to 1.1μm (both near infrared). Thus four different images of the same scene are obtainable, each depicting the scene in slightly different fashion. We can choose whichever view or combination of views will serve a given task best.

The data transmitted by ERTS is collected on tape at the ground and, after some "cleaning up", may be used as taped data or made into imagery in a manner analogous to that described earlier for scanner and radar data. An image produced from, say, only green light on black and white film will be a black and white image. But the tonal signature of a particular feature on a green light image will differ from the tonal signature of the same feature as depicted by infrared, and therefore give the viewer different or additional information (fig. 5-4).

Figure 5-4: Western Colorado as viewed by ERTS-1. The left view is a green light (band 4) image, the right is infrared (band 7). Field patterns in

The various scanner bands can be presented on photographic film in any desired assortment of colors. However, one of the most meaningful is false-color infrared imagery. ERTS data from the green, red, and infrared channels are each presented as a separate color image; then the three are combined at the Data Processing Center to create a color IR image (plate 12). Since all three data sets were collected simultaneously by the same sensor system, registry is not a significant problem.

While this sensor is a member of the same family as the optical mechanical scanners discussed earlier, a spaceborne scanner faces requirements different from those faced by airborne systems.

the valleys of the Colorado and the Uncompaghre Rivers contrast most
sharply. (NASA photos)

The nature of the targets of the ERTS scanner and the data quality
requested by the users increase the complexity of these demands.
The resolution capability of unclassified (militarily) scanners only a
year or two ago, for example, would not have been satisfactory for
sensing from 500 miles up. Indeed, there was widespread expression
of doubt that ERTS scanner data would be of much value for this
reason. The preliminary investigations that have utilized ERTS-1
data have shown that this factor is far less of a problem than
expected (ERTS Symposium, 1973). The quality of the imagery
appears at a glance to rival astronaut photography, a circumstance
not imagined possible; the hoped-for resolution has been exceeded.

Most surprising, perhaps, are some of the uses discovered for the data in tape form, a further reflection of the data quality since there is more detail in the tapes than emerges in an image format presentation. Plaudits for much of the success with taped data must go to the data processors who have found ways to dress up, enhance, or modify the raw data fresh from the sensors. Researchers have devised ingenious methods to increase the data's utility beyond any previous expectations. But the major credit must go to the scanner itself, since it is responsible for the basic data quality.

The other sensor system of ERTS-1 consists of three television cameras. One is for sensing in the green light band, one in the red, and one in the infrared as with the first three bands of the multispectral scanner. One capability of this system stems from the fact that green wavelengths provide maximum water penetration, while red does so less well and infrared poorly if at all. Subsurface features in shallow water thus should be discernible from one data set, but margins of water bodies will be more sharply defined by another. Each camera has its own alignment with the earth's surface, and although all three are aimed at the same area, the images are not identical. The problem of registry, not encountered with the scanner, must be solved by electronic processing on the ground when combined images or data are desired.

These television cameras, called return beam vidicon (RBV) cameras, represent an advanced level of sophistication also. While the standard U.S. television camera picture frame is based on about 500 lines, these cameras utilize over 4000. They are perhaps the highest resolution TV cameras to fly in space so far (Strickland, 1972). As with the scanner, the data collected by the cameras originally was transmissable as real-time data or could be stored onboard for subsequent transmission when in range.

The initial pictures from these cameras obtained within two days of the July launch served notice that the sensor systems had some pleasant surprises in store; the RBV cameras were performing even better than expected. Then in October a problem developed with one of the power control switches of the satellite which could have caused damage to the vidicons, and for their own protection they were shut down. One of the video tape recorders onboard also failed; while the other one gave no indication of a problem, the RBV camera system was shut down indefinitely or until circumstances changed substantially. The scanner system continued to function, however, and to supply data for some 300 experiments that were funded to investigate the utility of ERTS data. More of the experiments were dependent upon scanner data than RBV data, so the shutting down of the RBV cameras was less of a blow than shutting down the scanner would have been. Still later, the tape recorder which was serving the scanner suffered a partial mechanical failure, essentially limiting the satellite's capabilities to real-time data, transmitted when it was near one of the four receiving stations.

Communications Role

In addition to its primary mission of depicting the character of the earth's surface, ERTS-1 has a less well known role as a real-time communications satellite. As mentioned earlier, ground-based sensor data is needed from areas not served by manned observation stations, especially in the empty reaches of some of the oceans or in rugged mountains. Parts of the air ocean also are too remote from ground receiving stations for practical data transmission from instrumented balloons. Instrument packages in such areas equipped with radio transmitters requiring so little power that they could transmit continuously could solve the problem if a receiver system could be brought in proximity repeatedly. An orbiting satellite is the obvious platform for such a system.

ERTS-1 receives and retransmits information from data collection platforms mounted on buoys (water and air temperature and motion), from ground-based data collection platforms (weather, hydrologic, and seismic data), and from constant level balloons (temperature, pressure and winds) in remote areas. The 100 or so experimental earth data collection platforms probably will be increased in number as their worth is proved, since the ERTS-1 Data Collection System can accommodate ten times that number. Although the system can serve only instrument stations that are in ERTS's proximity at a time when the satellite's transmissions can be received by the ground stations, a vast area is encompassed in the various swaths covered by ERTS's orbit. Each data-collection platform transmits a "burst" message (time compressed) every three minutes. This transmission frequency is related to the schedule of satellite overflights to assure satellite relay of at least one burst message every 12 hours. The ability to obtain real-time data from many such remote locations is a powerful new tool for agencies whose effective service is highly dependent upon up-to-the-minute widespread data coverage.

ERTS Overview

How useful is the data being returned from ERTS-1? What can be done with this information? An early attempt to answer such questions was made at a meeting in March of 1973 (ERTS Symposium, 1973) where investigators who had begun projects using ERTS data compared first results. It must be emphasized that the results reported were progress reports of research underway, not final results. Additionally, a tendency to emphasize the positive was understandably evident, so that the following excerpts are bright with promise. Even if only half the prospects become reality, however, they show the increased recognition of the value of satellite-generated data.

The identification of selected crops was achieved by some researchers with accuracy figures of 80 percent or higher. In an area of intensive agriculture where small fields—some as small as five acres—predominate, fields in crops, ploughed fields, and fallow fields were distinguishable from one another. Structural features of the earth's crust occasionally proved more visible than on some of the astronaut photography earlier used to study the same area. Vegetation cover change even in sparsely vegetated rangelands was found to be observable, indicating a potential for estimating carrying capacity. Changes in the size of water bodies were observable, implying changes in the water table as well as the overall water resource situation. Contrasting signatures in predominantly hardwood and pine forests were identifiable, indicating forest-type mapping capabilities. Differences in the appearance of broad lithologic categories provided the basis for creating an areal geology map. Hail- and wind-damaged areas in large fields of crops were discernible and mappable. Gross soil differences such as clay-rich as opposed to sandy soil provided visibly different signatures. The better haze penetration, the superior definition of water bodies by the IR band, the ability to use tomorrow's imagery if today's is too cloudy, the improved angular relationships of the imagery, and radiometric fidelity are factors which are causing cartographers to view such imagery as the basis for vastly improved mapping capability. Current maps of the Antarctic and of Brazil have already been shown by ERTS-1 imagery to have location inaccuracies.

Perhaps most intriguing is what can be seen with data manipulation. It is possible to display portions of the taped data on a TV viewer and blow up the scale until the image loses definition. Just short of this point (which varies with the target of interest) objects of modest size can be identified, using foreknowledge. The Washington monument—and its shadow—in Washington, D.C., can be so located. To be sure, it is a poor view of that monument, but it was made not with a camera but with a scanner some 500 miles away. Such examples make realistic comments of resolution capabilities difficult to phrase.

Considering that ERTS-1 is only an experimental prototype of future operational earth resources satellites, the potential it has indicated points clearly to one conclusion. Man simply will not settle for earth information collection procedures in the future that do not include satellite-generated data.

Skylab

In May 1973 another satellite, called Skylab, was launched by the United States. To provide information about the earth and its resources was only one of Skylab's tasks and is about the only similarity between Skylab and ERTS. Skylab is a manned satellite, which accounts for the most obvious difference between the two—size (fig. 5-5). ERTS-1 weighs about one ton, Skylab about 100 tons. Skylab is only about half as far above the earth as ERTS-1 and does not have a polar orbit; and data collected from Skylab was brought back to earth by astronauts rather than being sent back by telemetry.

The differences between the missions of these two satellites will become evident in the discussion of Skylab which follows, but a comment on relatedness may be appropriate at this point. Study of the earth by spaceborne remote sensors onboard Skylab may seem duplicatory of the ERTS-1 mission, but the roles of the two satellites are in fact complementary. ERTS-1, remarkable though it is, carries a very limited sensor package; the sensors on Skylab that were used to study the earth are both more numerous and more complex. This increased capability raises the question of how to insure that such instruments are most effectively utilized. Skylab has given us a hint by partially answering another question, "Would some remote sensing from space benefit from man's presence there?"

The answer to this second question, for the initial Skylab mission at least, was provided in unexpectedly dramatic fashion with the repair of the crippled satellite by the astronauts who followed it into orbit later in a separate vehicle. The reader will recall that the mission seemed doomed at the outset when Skylab was

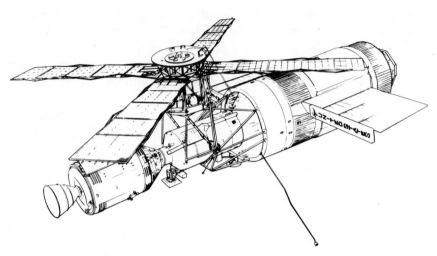

Figure 5-5: Skylab. The CSM "taxi" is lower left, docked at the axial port of the MDA, above which is one of the solar array systems. The other system, projecting from the sides of the workshop, was damaged during launch, and the panel on one side was lost. (Courtesy of Martin Marietta Aerospace)

damaged while being put into orbit. Not only was repair facilitated by on-site description of the damage and effected in orbit by the astronauts, but subsequent additional problems also were solvable because of their presence. To be sure, the astronauts had more important questions to answer than whether they could accomplish repairs. Still, one could hardly question seriously the value of their presence on this mission.

The Mission Profile

It should be made clear at the outset that the Skylab effort is a little different from anything that we have done in space before. It is the totality that is most different, however; a little reflection on the component parts of the mission will bring to mind similar earlier activities in space. The Skylab vehicle, complete with instruments and living accommodations, went up unmanned, and the men achieved rendezvous with it later. We had accomplished numerous space rendezvous of one sort or another before, but we had not sent up a space station as such. For that is what we have in Skylab—a form of space station to be visited at least three times. Space-walks, or extravehicular activity (EVA), were used routinely on Apollo to recover film and experiment samples from the Service Module. EVA also was a routine activity on Skylab, although some of the first operations were neither routine nor planned. There have beeen several attempts to test man's ability to remain in space safely for protracted periods, and the crew's ability to do work in space has been tested as well. But the limitations of room within the earlier vehicles placed serious constraints on the kind of activities which crews could perform. In Skylab the men were approaching a daily work routine like that in a laboratory on earth more closely than has been previously attempted. It is stretching things a bit further to compare their living conditions to those on earth, but such conditions were approached more closely in Skylab than ever before.

The Skylab vehicle was put into orbit by a Saturn V launch vehicle. The combination of the two defies adequate description. Its length of over 330 feet approximates the height of a 33-story building, but what we are talking about doesn't resemble any kind of building in the slightest. To take note of the combined weight as being over six million pounds does little better—what does the average person know of that weighs six million pounds? Even the indescribable fury of the engines at launch seemed inadequate to lift such a weight off the pad, let alone make it fly. But up it went and it flew—or more properly, orbited. Of course, all but some 200,000 pounds of the total weight was used up just putting it there.

The day after the launch of the Skylab vehicle, the three men to occupy it were to have gone up much in the fashion of other astronauts. It was only after a delay of several days, however, that their "space taxi", a command and service module (CSM) similar to those of the Apollo program, was put into orbit by a Saturn IB. While dwarfed by the preceding pair, the combination is still impressive—over 220 feet tall and weighing almost 1,300,000 pounds. After insertion by the Saturn into an interim orbit lower than that of Skylab, the CSM climbed to Skylab's altitude and achieved rendezvous by its own propulsion system.

The first crew remained at Skylab 28 days as originally planned, despite the problems. Although some of their time was taken up by activities that were not in the original mission plan, much of the solar research, the evaluation of the long-term habitability of the Skylab, and the variety of medical research experiments planned were carried out. At the end of their stay, the crew prepared Skylab for storage in orbit, reentered the CSM which brought them, and returned to earth with the non-transmissible data such as photographs and specimens. Two more crews visited

Skylab, each remaining substantially longer than the preceding one until all previous records were broken. During these longer visits the earth resources remote sensing experiments were a major activity.

Skylab and its crews were launched from Kennedy Space Center. The Skylab orbit is inclined 50° from the earth's equator; on its northward sweep it thus reaches 50°N. latitude before it starts south again, and it reaches 50°S. latitude at the opposite extreme. This orbit takes Skylab farther north and south than any U.S. manned spacecraft has gone, and allows it to pass over the majority of the densely settled parts of the world. Because of its size and its orbit, Skylab is visible from earth. At an altitude of about 270 miles and moving at a speed of five miles per second, it may be seen for nearly ten minutes at sunrise or sunset moving northeastward or southeastward. The orbit period is about 90 minutes, or about 16 revolutions per day. It is visible from every part of the United States (except Alaska) at one time or another, weather permitting.

The Skylab Vehicle

The total vehicle exclusive of the CSM taxi is referred to by NASA as the Saturn Workshop. It is composed of several major components, principally the Apollo Telescope Mount, (ATM), the Multiple Docking Adapter (MDA), the Airlock Module (AM), and the Orbital Workshop (OWS) (fig. 5-6). They were joined together in the order of listing at launch with the Telescope Mount at the top. After insertion into orbit the telescope complex with its extensive solar panel array rotated 90 degrees to a position beside the Multiple Docking Adapter (see fig. 5-5). The MDA, like most components, serves several purposes. It contains many of the sensors used to study the earth from Skylab. Its fundamental role, though, is as the docking station for the CSM which brings the crews to Skylab. Normally the CSM docks at the axial port, as shown in the illustration; an alternate radial port also is available, however. Other activities for which the MDA is equipped include materials processing in space, such as working with molten metal flow characteristics under zero gravity and vacuum conditions; operation of the earth resources sensors; and operation of the Apollo Telescope Mount.

The ATM is the heart of one of the four principal objectives of the Skylab effort—to study the sun with sophisticated equipment from a vantage point beyond the earth's atmosphere. Before the advent of satellites we were forced to study the sun through the atmosphere, although studies from high mountains and the employment of airborne sensor platforms such as balloons sought to reduce its effect. Some experiments in solar astronomy have made use in limited fashion of our recently acquired ability to pass beyond the atmosphere, but nothing approaching Skylab's battery of equipment has been put in space along with men to operate it for protracted time periods.

The sun is the principal source of energy on the earth. Were its output of energy to fail, the earth would quickly become dark, lifeless, and unimaginably cold. Interaction between solar radiation and our atmosphere is the basic cause of our weather, and variations in solar radiation undoubtedly help cause climatic change. There is much about this and many other earth-sun interactions that we do not understand, and a primary reason is that our information base is inadequate. The coronagraph, the UV spectrograph, the coronal spectroheliograph, the two X-ray telescopes, the polychromator/spectroheliometer, and the telescope cameras which are parts of the ATM provided an opportunity to study the sun as never before and thereby improve that information base.

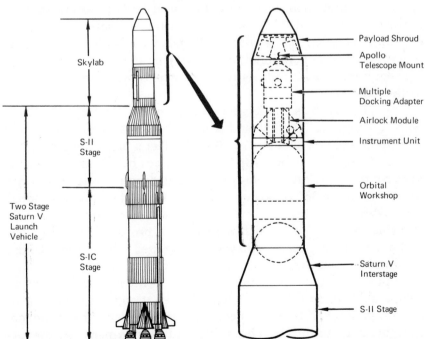

Figure 5-6: Skylab in launch configuration. The most obvious difference in appearance from the operational configuration is that the solar arrays are here retracted and folded away beneath a protective launch shroud. (Courtesy of Martin Marietta Aerospace)

Much of the information obtained with the ATM was recorded on film cartridges which had to be changed many times. Since arrangements for changing them from within Skylab were impractical, the crew accomplished this task by EVA. The airlock module, adjacent to the MDA, provided the means for leaving and reentering Skylab for EVA. A special hatch, an airlock compartment, and EVA support facilities such as umbilical equipment are found in the AM along with some additional experimental and operational equipment for the space station. The Instrument Unit is a small instrument-crammed segment whose most important functions were to guide the vehicle during launch and orbit achievement, and to control vehicle stability during the first few hours of Skylab's existence in orbit.

The only other major component of Skylab is the Orbital Workshop. It is here that the crews ate, slept, and carried on the housekeeping activities necessary to life. The facilities for these basic operations are only part of the equipment complement of the OWS, however. A second major objective of the Skylab mission was to study what effect living and working for protracted periods in the space environment has on man, and most such experiments were performed in the OWS. Additionally, many non-solar astronomical experiments such as stellar and galactic studies as well as navigational and operational activities were carried out from Skylab. The equipment and facilities for these experiments are also located in the OWS.

The OWS is the largest component of the Skylab vehicle, with a length of 48 feet and a diameter of about 21 feet. The workroom volume of about 10,600 cubic feet is almost ten times that of the MDA, the next most spacious unit, but hardly as big as is implied by the occasional analogy to a two-bedroom house. Still, when compared to the 360 or so cubic feet of a CSM it seems vast indeed. The OWS dominates the Skylab vehicle all the more because of the two solar array panels which were to have projected from its sides. Although much more compactly grouped than the striking windmill-shaped arrangement attached to the ATM, they would have presented the same amount of area to the sun and collected a comparable amount of solar energy. When it appeared that the use of both panels had been lost because of the accident, the entire mission became doubtful. The successful deployment of the one remaining panel achieved during EVA gave Skylab three-fourths of the planned-for energy, enough to carry out most of the experiments.

There are too many additional experiments, types of equipment, and activities associated with simply keeping an entity as complex as Skylab functioning properly and on course to attempt a complete survey in this presentation. No effort will be made, for example, to cover the third major mission objective, space technology studies concerned with manuevering in space, materials degradation, and repair techniques. The last major mission objective, remote sensing of the earth, does deserve special consideration here.

EREP

The Earth Resources Experiment Package utilizes six remote sensing systems: two camera systems, a spectrometer, a 13-band optical scanner, and two microwave systems. Collectively they have the capability of sensing over a substantial portion of the electromagnetic spectrum, including the visible and portions of the infrared and microwave regions. Experiments in agriculture, forestry, oceanography, hydrology, geology, meteorology, ecology, and geography have been devised to use selected parts of the data EREP collected. Many of the experiments are extensions of research mentioned in previous sections. EREP provides opportunities to investigate the potential utility of alternate forms of data for a given task, to seek further results from the advanced development level of its sensors, and to experiment with sensors oriented in different attitudes or used in different modes. Data acquisition procedures were modified repeatedly and substantially after launch by the crew members in the sensor platform.

The two EREP camera systems of Skylab are altogether different from the RBVs of ERTS—they are not TV cameras. They utilize high resolution photographic film that provides superior imagery. The camera systems operate on manual control, photographing selected targets under desired conditions with current information supplied from the ground for guidance. One system is really a composite of six separate cameras, each of which can be mated to a variety of film/filter combinations. From this system 70mm imagery is now available in several varieties such as black and white, color, black and white IR, and color IR (plate 13). The other EREP camera, the earth terrain camera, is a single camera with a longer focal length and ground coverage of substantially less area. It can utilize several film types, but its purpose is to provide very high resolution support imagery.

The IR spectrometer can be used by an astronaut to track selected ground targets and keep the sensor aimed at a given small target long enough to obtain a representative radiation signal from it. This capability permits more accurate estimation of the atmospheric effect on terrestrial radiation; such information can be used to assess the accuracy of data obtained by other spaceborne sensors. Data simultaneously collected from aircraft within the atmosphere and at the ground site itself will provide the basis for comparison. It is difficult to overemphasize the importance of this experiment since signals received by most sensors in space are affected to some extent, as yet imperfectly known, by the atmosphere.

The EREP multispectral scanner resembles the one on ERTS in operation and purpose but there is a notable difference. Instead of collecting four sets of data in four spectral bands, the EREP scanner utilizes 13 bands, half in the visible and half in the IR region. The more narrowly defined bands pinpoint more accurately the wavelengths which provide the most useful information for different tasks.

The remaining two EREP sensors both utilize energy in the microwave region. In general, they seek to establish the utility of microwave sensors from orbital altitudes for earth resources studies through microwave radiometry, scatterometry, and altitude sounding. Both active and passive systems are involved, in an effort to achieve results comparable to those obtained from similar aircraft systems.

Skylab, like ERTS, is part of an ongoing NASA experimental and research program whose results are not intended for operational application to current everyday problems. The results from both, however, should give us a good idea of the role that space-generated data can be expected to play in the solution of future everyday problems. The nature of the truly operational satellites to come and the role to be played by manned satellites will largely be determined by the results of this program.

Tomorrow

This book was conceived during the spring of the year that the first United States earth resources satellite was to be launched. Since the writing began, ERTS-1 has been launched, its sensor systems have returned data, and research employing the data is underway. Skylab also has been launched, and before the book was published its mission was completed. Events in this field are moving at a rapid pace.

As the launch of ERTS-1 approached, widely varying opinions about it were expressed orally and in print. Some of the things written about the satellite and its planned activities have proved to be incorrect. Events do occur which require or make desirable changes, even in meticulously made plans. Because of both the rate at which things are happening and the certainty that some of what is written here about forthcoming events will prove inaccurate, for anyone to attempt a look into the future requires something approaching temerity. Some sort of look, even if only a glance, seems called for, however, since recent events seem increasingly to point in a similar direction.

Remote sensing from space must surely increase in importance, although the rate at which it will happen and the degree of its importance to remote sensing in general are dependent upon too many variables for sensible conjecture now. The timeliness of such data, the continuous sequential coverage that it affords, the uniqueness of the view provided, the quality of the data, and the potential it promises constitute a development too valuable to waste through limited use. An expensive development it has been, to be sure, but this very fact is one of the strongest arguments for implementation. One does not spend a king's ransom to develop a tool to the point at which it can begin to pay for itself and then put it aside, especially one with such wide application.

It is likely that operational earth resources satellites will become at least as familiar a phenomenon as weather and communications satellites. The chances for their becoming even more familiar are very good because of the breadth of their application and the nature of their concerns. It appears that they can monitor crops, rangelands, and forests for symptoms of stress or for harvest estimates, and provide input for hydrologic studies and mineral exploration. They can serve as an additional data source for studying the nature and extent of change in land use and urbanization. Air and water pollution monitoring are likely to benefit also from satellite surveillance. Weather satellites will continue to improve their contribution until far more accurate and longer range forecasts are routine, and communications satellites (which now carry over half the international telephone, telegraph, and television traffic) and navigational satellites will assume additional roles. To what extent these many services will be provided by multipurpose satellites remains to be seen, but it is clear that there will be many more satellites than most of us now envision.

The further development of these tools and the creation of operational models of earth resources satellites will continue to be expensive. Even though the expense is confidently expected to be offset many times by their contribution, ways to reduce the original expense should still be explored. Actually, such efforts have been underway almost from the beginning, but the most impressive effort of this type may lie ahead. A large proportion of the cost of all satellites is represented by the equipment necessary for putting them into orbit, notably the booster, which typically can be used only once. The space shuttle program will seek ways to alter this wasteful situation.

Space Shuttles

To date, the vehicles man has used for space flight and those used for flight in the earth's atmosphere have, by and large, been distinctly separate entities. Heavier-than-air craft employ wings in conjunction with a power source to support the vehicle while it is airborne. Even the helicopter works this way, since its rotor blades are themselves a form of wing. Since in space there is no atmosphere to support an airfoil, wings have no utility for space flight; here the velocity supplied by rocket thrust must both combat the effect of gravity and supply a directing force. Ultimately, vehicles may be devised that are equally at home in the two environments, but such a craft poses complex problems such as reentry into the atmosphere. The velocity required to keep a spacecraft in control outside the atmosphere becomes destructive when the vehicle encounters the atmosphere.

The major economic problem in question at the moment has to do with the equipment required to put the vehicle into space. Enormous amounts of fuel are expended in a launch; the fuel containers, the equipment necessary for properly supplying it to the engines, and the rocket engines themselves constitute the bulk of a total space vehicle before launch. Between 30 and 40 times the weight of the Skylab vehicle or the CSM "taxi" is required just to put them into orbit. Once in orbit this launch equipment is of no further use to the mission, and after separation it is written off because of the impracticality of recovery. In fact, one flight is all that most space equipment has been used for, even recovered items. Finding ways to minimize the waste of one-time-only use would do much to reduce the costliness of space activities.

The name "space shuttle" implies a vehicle which will shuttle back and forth between the earth and space *repeatedly*. In order to do so, the vehicle will need the compatibility with both environments mentioned earlier, i.e., be a space ship during the launch and while in space and an airship at the end of the mission. The shuttle presently visualized is a three-stage composite: a pair of solid propellant rockets, a large liquid propellant tank, and the orbiter vehicle. The most power is required during launch when the total assemblage has the role of a rocketship; at this time the attached solid propellant rockets add their thrust to the orbiter engines burning liquid fuel. At an altitude of some 25 miles, the solid propellant rockets are ejected, and descend by parachute to the sea where they are recovered for subsequent reuse. The orbiter then achieves orbit insertion through use of its own engines, still fueled from the large tank. Once in orbit this tank also is jettisoned, although its recovery is not attempted. Since it is essentially only a fuel container, its loss is in no way comparable to that of present launch vehicles which include, among other things, rocket engines. The final stage, the orbiter vehicle itself, still will have the requisite fuel supply and engine power for accomplishing its purpose and returning to earth. Fuel requirements for limited activity in space and for return are modest.

At least two characteristics will make the orbiter different from any previous space vehicle launched. For reentry the entire vehicle will have a protective coating, analogous to the heat shields on the base of the vehicles which have returned astronauts to date; and the vehicle will have wings for descent through the atmosphere to a landing field. It is expected that these characteristics will make the orbiter usable again and again.

Such a space shuttle could replace one-time-only launch vehicles for most tasks. It would *be* the launch vehicle for many of the kinds of satellites mentioned earlier. Planned for an average crew of four, a space shuttle could take technicians to orbiting satellites for repairing or modifying them. Or the satellites themselves could be returned to earth by the orbiter vehicle, thereby reducing costs even further. Conceivably, more sophisticated shuttles could even go commercial. If the plans for passenger accommodations become a reality, you would not have to be an astronaut to make use of space shuttles.

The exact nature of the space shuttle and its capabilities may turn out to be substantially different from the view offered here, but the important point is that the concept marks a more determined trend toward economy and practicality in space ventures. One cannot start a space program with these as dominant constraints and achieve the success that we have experienced. That we can think in such terms now shows that we have advanced beyond the purely experimental stage of space flight.

Chapter 6

Pro and Con

For some 100 pages now, in extolling the virtues of remote sensing, we have alluded persistently to the potential for great benefit which this new tool offers. Occasionally we have sounded a precautionary note or voiced a mild remonstrance against premature enthusiasm, but the overriding attitude clearly has been positive. There are, however, negative aspects to remote sensing which we must consider if we are to achieve a representative overview. Some of them have not yet been recognized by the man in the street, so recent is the development of the technique and so limited the understanding of its capabilities. Other aspects are seemingly less difficult to understand and have been the subject of spirited debate for several years. We moved past the "gee whiz" stage of remote sensing some time ago, and some people have been questioning certain aspects of the development programs. Predictably, a principal focus of such questioning has been the factor of cost.

Cost

This topic is commonly replete with pitfalls. It is all the more so for remote sensing because of the developmental stage of the technique. Even the most basic data, such as actual expenditures, is difficult to pin down since some of the research advances made have been parts of more inclusive efforts. Moreover, it is unrealistic to consider the cost of this kind of development without also considering benefits, and benefits at this stage of the game can only be prognostications. In many instances, remote sensing is likely to be most valuable in a complementary role, and assessing proportional contributions is a challenge. These are some of the reasons that the limited treatment of cost here will lean heavily upon comparisons.

There is no way realistically to depict the development of remote sensing as inexpensive. It probably will pay for itself many times over eventually, but the expenditures to date have been substantial. Those parts of the space program which have concerned remote sensing are impressive in themselves. Although remote sensing activities are only part of Skylab's mission, some critics tend to assign the total program's cost of over $2.5 billion to the remote sensing debit sheet. Skylab need not be mentioned at all, however; the $100 to $200 million (depending upon how many items are counted) allotted to the ERTS program, which is clearly a remote sensing experiment, provides a respectable enough example. Remote sensing research and development has not been dependent upon developments in space, though. A vast but little-heralded aircraft program, mounted early in the development sequence, continues to make major contributions to remote sensing development, and it does not run for free. Innumerable grants for research have been given to universities, government agencies, and commercial firms for experimentation and the development of hardware and software for over a decade; collectively these items are also a large investment. How can such expenditures be justified?

A sense of perspective, without which there is little hope for objective judgment, is required for considering this question. Dollar amounts alone are misleading. We are in an era of routine expenditures of prodigious sums; where several decades ago we spent million and tens of millions, today we spend hundreds of millions and billions. It is the dimensions of the problems to be solved and the returns to be realized that must be considered. They must be appropriate to the expenditure in order for it to be justifiable. Do the problems remote sensing may help to solve have such dimensions? Agriculture has been the focus of much remote sensing research and has been discussed earlier; let us consider it in the above context, from the standpoint of a business proposition instead of a necessity for life.

Over the years, agriculture has become a very big business. The agricultural exports of the United States alone are worth on the order of $10 billion per year, and we export only a fraction of our total produce. If we shift our focus to world agriculture, hundreds of billions of dollars are involved. Agriculture thus appears to have dimensions such that large expenditures for the solution of its problems would be appropriate.

Much research in weather and the development of weather satellite capabilities has in fact been carried on with agriculture in mind. Directly weather-caused crop loss in this country in recent years has been averaging between $1 and $2 billion annually (National Academy of Sciences, 1970); a 5 percent reduction in that loss because of improved forecasting should result in a $100 million savings. Figures of this sort often are difficult to relate to reality, but the events of the autumn, winter, and spring of 1972–1973 are fresh enough in mind to make it easier. The crops left unharvested in the Midwest fields caught by a wetter than average autumn, the livestock losses due to cold and unusually heavy snows in the West, and the flood damage to fields in the spring easily account for losses of many millions of dollars. With improved longer range forecasting, such losses could be mitigated.

An estimated $7.5 billion in annual crop loss caused by insects and disease is another problem the United States faces (National Academy of Sciences, 1970). Only a 1 percent loss reduction here would net a return of $75 million. In recent years about $3 million has been spent annually on detection of crop disease and insect problems. It is believed that with the help of ERTS or its successors this figure could be cut to $1 million. The U.S. Department of Agriculture estimated several years ago that a $5 return for each $1 spent on ERTS would be possible with the full employment of ERTS data (Aviation Week & Space Technology, 1969).

The problems of the nation's forests also are large enough to justify the application of remote sensing. Forest fire losses in the United States have been averaging between $300 and $500 million annually in recent years and forest losses due to insects amounted to almost $600 million in 1965 (National Academy of Sciences, 1970). The capabilities of thermal infrared scanners in forest fire detection and suppression were cited earlier, as was the detection of stress in trees from disease or insect pests by color IR imagery.

For a variety of reasons, airphoto coverage of large areas of the United States is needed, and this coverage must be updated periodically. In the past it has been acquired piecemeal over a number of years and the resultant imagery included hundreds of thousands of individual photos. In the future we will require such coverage more often, and the scales obtainable on ERTS imagery by data manipulation appear to be satisfactory for many purposes. It is difficult to make valid cost comparisons between aircraft photography and satellite imagery, in part because no subsequent set of satellite imagery costs quite as much as the first—the satellite had to be launched to get the first set. Cost calculations on this point are further complicated by the fact that costs associated with experimental satellites or sensor systems are not comparable to operational situations. But from the standpoint of different types of dimensions, the volume of imagery involved, and the time to acquire it, satellite imagery involves fewer problems. Between a half million and one and one-half million airphotos (depending upon scale) are needed to cover the United States; four to five hundred ERTS images will do it and total coverage is available in days rather than years. The opportunity to achieve such a reduction in these dimensions would seem to warrant a substantial expenditure also, especially since some experts insist it is cheaper this way.

A number of additional examples of application will occur to the reader on reflection over the material in previous chapters. While many of these individually may not have large dimensions, collectively they certainly do. This point could be elaborated at great length. Satellite remote sensor coverage is world-wide areally and continuous temporally; and we have tried to show the diversified applicability of remote sensor data in general. The answer to the question of whether the problems being investigated have appropriate dimensions must surely be yes.

There are too many aspects of the cost question for a complete review in a volume this brief, but two additional points will be mentioned. First, remote sensing is not an inexpensive technique. The kinds of information being sought are unusual and the equipment, complicated to begin with, is becoming more and more sophisticated. Secondly, there will continue to be experimental development for some time to come. Although there are operational examples, the realization of remote sensing's full potential will require much additional and costly research. This expense was foreseen when the original investment was made. Whether or not the decision to make the investment was wise cannot properly be answered until remote sensing has reached the fully operational stage and the results can be compared to the cost.

The State of the Art

The Basis for Confusion

If it is difficult to come to grips with an appraisal of what remote sensing developments cost, in some ways it is even more of a problem to assess its development status. For one thing, developments are occurring at such a pace that merely attempting to keep abreast of them proves frustrating. Some of the capabilities described in this volume are changing as it is being written, and most will have changed in some degree within a short time after its publication. Then too, progress on the several fronts of remote sensing is uneven. With some sensor systems diversification has enhanced capabilities, while with others it has delayed advances by fragmenting the effort.

Perhaps most distressing was the oversell period of several years ago which left a legacy that is still with us. The exciting potentials hinted at by preliminary investigations inspired sweeping claims, some of which have not been borne out by subsequent studies. In some instances the sample size in the original investigation proved to be inadequate; in others the claims were simply premature. Remote sensing is not new anymore, but it is still very young. Those expecting a performance typical of a mature technique are apt to be disappointed and thus become disenchanted with the whole thing. Such disillusionment and antagonism can only hinder the development of remote sensing.

The problem of perspective is further complicated; there are examples of extant applied research which either represent actual operational use or are so close to it that distinction becomes questionable. Aside from obvious cases, such as the National Weather Service, there are examples of use in limited situations by other government agencies such as the U.S. Forest Service. Furthermore, commercial firms are becoming involved in remote sensing: there are corporations who sell expertise as well as those who manufacture and sell the equipment. One of the more prestigious (by virtue of its personnel roster) new expertise firms markets its services on a world-wide basis and has contracted with several nations to provide resource surveys and mapping data. The existence of such activities makes it difficult to visualize remote sensing as being immature, but such examples are recent and few. The widespread operational use of the technique lies ahead.

Problems

The relative youth of remote sensing is reflected in such things as uncertainty regarding the full meaning of some of the data being collected, collection of data that was not sought, failure to collect data that was sought, and outright equipment failure. Many of the pieces of remote sensing equipment in use have been prototypes, even though not first-stage models, and their operation has not always been consistent. A rather misleading impression regarding equipment, incidentally, stems from the number and diversity of remote sensing research activities over the past few years. A great deal of mileage has been gotten out of relatively few pieces of equipment; there just has not been that much available. The current situation of ERTS-1 is a prime example: several hundred research projects have been utilizing data from one sensor system in one satellite. The picture is changing as more and better equipment becomes available daily. We are some time away from a situation in which any piece of necessary equipment can be expected to be available, however.

One of the more frustrating problems for those who would like to experiment with or teach remote sensing has been data acquisition. The limited number of sensor systems and their cost or limited availability has made imagery samples hard to come by. In the case of space-generated data, the matter has been further complicated by the small number of sources from which we could expect to obtain data. It appears that both the federal government agencies involved and the manufacturers of commercially available equipment were caught by surprise at the size of the demand. There are other possible explanations for the data availability situation also, but in fairness, a view from the distributor's seat deserves consideration. There is a tremendous volume of material to be processed, catalogued, and stored. Then too, quality variations must be noted, since they are critical to users. Retrieval and reproduction systems are essentials also, as well as distribution considerations. Not exclusively, but especially for space-generated data, spatial and temporal information must be correlated with the data since requests are unlikely to be couched in terms of orbits and frame numbers. In short, a colossal amount of organization and effort is necessary to produce the specific pieces of imagery people request at the times they request them. As of this writing, that status has not been achieved; we are still in the setting-up period. Much data can be obtained through the existing channels, however (see Appendix).

Since some data is *unobtainable*, the point of security classification should at least be raised in passing. What sorts of data and imagery exist that have been acquired by classified remote sensing activities will remain a question for most of us. Occasionally accidental (or intentional) "leaks" of information find their way into the media and we get a glimpse of greater potentials. Most of these leaks have to do with resolution capability* and they tend to emphasize the spectacular: "An object in orbit the size of a basketball can be detected from earth," or "An individual person on earth can be resolved by spaceborne sensors," or "The ability to photograph the Lunar Rover on the moon's surface from an Apollo spacecraft is the equivalent of being able to photograph a Volkswagen on earth from orbit"—and so on. One tends to doubt that in fact "military officers' ranks can be read from their shoulder insignia from space", yet it seems clear that classified remote sensing has capabilities distinctly superior to the unclassified.

This gap is unavoidable, and perhaps even desirable; but we are not debating the theory of governmental policy here. What is more disturbing is the possibility that some data or capabilities remain classified long past the point of need, which would result in a regrettable waste of effort. A very considerable amount of research is directed toward attempts to wrest from sensor returns far more than is immediately apparent from the basic data. While success in such efforts might be transferable to more sophisticated data and would make them more usable in time, there are tasks for which extra effort would be unnecessary with more sophisticated data. The researcher who has spent months or years seeking ways to circumvent a data inadequacy problem will not be amused to learn later that needlessly classified adequate data was available all along. Few researchers argue with appropriate security measures. There is a distinct feeling on the part of many, however, that a review of classification policies is needed.

One other problem associated with the state of the art is the practitioners of the art. There are several facets of this problem; two are mentioned here.

The number of people sufficiently acquainted with remote sensing to be able to employ it effectively is relatively small, although the ranks of the informed (or "infected", as one wag puts it) recently have swelled impressively. But for employment of this tool on the scales that have been discussed, especially in an operational framework, the numbers are disproportionately small. The distribution of informed personnel is also uneven from several standpoints. Although there have been efforts on the part of several agencies of the U.S. government (notably NASA and the Geological

---

*Actually, as has been implied before, resolution per se has ceased to be the primary consideration it was earlier. It has become increasingly evident that the utility of data should not be judged by this parameter, and that the parameter itself is difficult to define simply (Rosenberg, 1971).

Survey) to acquaint foreign nationals with the development of the technique, the great majority of nations remain ill informed. Some that could most use it are the least aware. Within the United States there are government agencies—federal, state, and local—that could be incorporating remote sensing into their planning and programs, but are not doing so effectively or at all. Additionally, the academic disciplines have uneven representation in the research on application and development of remote sensing. Although most fields are represented by at least some effort, there are cases in which it has been modest indeed.

This situation is a reflection of the relative youth of the technique, but it represents a potential stumbling block for the implementation of remote sensing on a large scale. Even if the other problems referred to in this section are solved, a shortage of informed personnel could still be an obstacle. As the reader may now better appreciate, mastering the concepts of remote sensing at a level appropriate for actual involvement will take a little time. And so will establishing programs for making use of the tool.

The other facet of the problem with practitioners is a very human one—overenthusiasm. There are several ways this phenomenon manifests itself. The use of a sensor inappropriate for the task at hand is one. Another is the use of remote sensing where other techniques are adequate or even better. Then too, there is the temptation to try to do it all by remote sensing instead of using the technique as a complementary tool. The results of such actions are predictably unimpressive and they can be detrimental by providing the basis for valid objections. Hopefully they too represent just another reflection of the state of the art.

Conflicts

There are several negative aspects that have emerged during the development of remote sensing which could have been treated in one of the above sections, since either cost or growing pains has contributed to their controversial nature. The somewhat more evident adversary component present in each, however, seems to deserve separate emphasis.

There already are a few instances in which private enterprise has believed itself to be in competition with federal government activities in remote sensing. In spite of the foregoing comments regarding the state of the art, there are a few enterprising organizations that do have the expertise and access to equipment which enable them to offer remote sensing services at a price. Potential customers include state agencies, foreign governments, and commercial firms which do not have the requisite facilities or skills and would like to see if this new tool can help solve their problems. A firm wishing to monitor the effectiveness of its pollution control facilities, for example, might employ remote sensing specialists for the task. But the investigation of such a problem also has appeal to federal agencies or university research groups whose study grants and data are both obtained from the government. Such occurrences have been relatively few and are to be expected, perhaps, in a period during which it is desirable to encourage experimentation that is application oriented. Were they to become widespread, however, the result could be serious and a rather unequal contest. The government has made some moves to avoid having this kind of thing become a problem; making the data returned from ERTS-1 available to the public is an example. ERTS-1 data is said to be presently as accessible to commercial concerns as to those carrying on government-sponsored research. For that matter, the data is just as accessible to you or me, at least in some forms. Although the delay in obtaining requested material may seem to belie this statement, it can reasonably be assigned to the stage of development of the distribution system. The intent is for equal opportunity.

A different sort of conflict arises from disagreement among the several agencies of the federal government which have an interest in the development of remote sensing from space on the way such development should proceed. It is no secret, for example, that for some time a difference of opinion has persisted concerning the relative importance of manned and unmanned satellite programs. As remote sensing from space has been shown to have increasing potential, the question has assumed greater significance. The cost of a manned satellite experiment is, of course, many times that of one which does not involve astronauts. Many feel that unmanned satellite research yields a greater return for a given expenditure.

As with so many apparently simple problems, a little investigation reveals complexities. For example, if more of the alloted funds are diverted to unmanned satellite research, the manned space program will be reduced, since the socioeconomic climate of the past several years has not been amenable to increasing the total allotment. There is probably a minimum funding level below which such a program cannot be effectively maintained, though there is disagreement about precisely where that level occurs. Aside from the various consequences to the national interest, which are arguable, such a decline in funding can alter the risk factor, which for manned flight is less arguable. The incorrect identification of that minimum funding level ultimately could result in the manned space program's being shelved. And yet, there are many questions about remote sensing of the earth that will be answered more effectively with assistance from manned satellites at our present stage of development.

The quality of the data returned from ERTS-1 and the preliminary success with its application are clear indicators of a far larger role for unmanned satellites in remote sensing. It is, in fact, likely that most remote sensing from space will be carried out with unmanned satellites as we move into the stage of operational capability. Manned satellites will have other roles to perform, doing tasks which unmanned systems cannot accomplish. Perhaps one result of ERTS-1's success will be a move away from an "either-or" posture on the question and a greater determination to advance both manned and unmanned satellite development.

A more basic question which has been raised is not what kind of space program we should have, but whether we should have one at all. Again, although the development of remote sensing has not been dependent upon the space program, the indications are that the tie will become increasingly close. The prominence of the question and its bearing on remote sensing seem to justify our considering it briefly here; more thorough vindication of our space effort can be found elsewhere (House Report 92-748, 1971). Most critics who raise the question seem to feel that the money we have spent on the space program might better have been spent in one of two areas: health, welfare, or other social programs; or ecology-oriented projects.

Some $38 billion was spent on space in the decade beginning 1961. During the same decade, however, about $300 billion more than that was spent on social programs. There is no question that the space money could have funded many additional social programs, but there is serious doubt that the money actually would have been so allocated if it had not been spent on the space effort. Public demand for an answer to the challenge of Sputnik was at least partially responsible for the government's raising and appropriating the funds to begin with. And for various reasons, health and welfare legislation did not have as high priority in the national budget as it presently has. The priority question is less important now, since we spend over twenty times as much on social programs as on space.

Remote sensing, and especially spaceborne sensors, may well become the most powerful tool we have for sound ecological management, as Chapters 3 and 4 point out. Arguing that space money might better have been spent on improving the environment is thus, to say the least, a formidable task. Not only will the information from satellite sensors help Americans with their own environmental problems (problems which might not otherwise be solvable), but the people and ecology of the entire world will benefit.

Responsible Use

The evolution of a new tool or capability substantially advanced beyond those in existence upsets whatever balance had been achieved with its predecessors. A major or minor readjustment takes place as man seeks to take advantage of the new opportunities and deal with the problems their development presents. Often it seems that his interest in the opportunities is substantially greater than his concern for the problems. Many potential opportunities offered by the development of remote sensing have been discussed in earlier chapters. What are some examples of the future problems it poses?

Legal

In part because of its newness, but notably because it presents a variety of problems not encountered before, remote sensing has raised legal questions which are not likely to be resolved soon. Shortly after the development of the airplane nations began to question the desirability of having aircraft from other nations roaming within or crossing over their borders, even in peacetime. Some degree of legal recognition emerged for the concept of sovereignty of airspace, the idea that the air above a nation was to be considered a part of its territorial extent. Some nations have become increasingly vigilant in seeking to maintain this sovereignty as airborne surveillance techniques have improved. Airspace is voluminous, especially with a nation of any size, and maintaining its sanctity is difficult and costly. It can be done fairly successfully, however, by those willing to pay for policing and enforcement equipment. Airborne remote sensing of one nation by another has led to serious incidents such as the Cuban missile crisis.

How high is the outer boundary of a nation's airspace?

There are several factors which make this question immaterial at present. Satellites have a minimum altitude on the order of 100 miles, are relatively tiny, and orbit at high velocities; detecting them is difficult and preventing their passage is only a theoretical possibility. Additionally, the increasing sophistication of sensor systems makes questionable the value of any attempt to establish a minimum altitude for satellites. It is likely that present sensing capabilities of military satellites are at least equal those of high-altitude aircraft of a few years back and will probably increase. At least with aircraft it was reasonable to restrict certain areas and to require detours around them. For a variety of reasons, detouring satellites is impractical; moreover, satellites cannot have half-orbits. Thus there are a number of intelligence satellites of various nationalities routinely surveying many parts of the earth (including the United States) without explicit legal authorization to do so.

Perhaps because most people do not know what satellites can "see" or are unaware of their existence, objections by the public to their spying have been relatively few and ineffectual. Weather satellites have not posed a problem because, in addition to their low resolution capability, the data has an obviously beneficial value and is available to anyone. As a trend toward satellite study of the earth's surface and earth resources began to develop, mixed reactions appeared. To some extent, objections to the idea were based on lack of understanding. This reaction is easy to understand for those who remember watching Sputnik; we did not know its purpose or capabilities, nor the intent of its creators. Other objections have a more problematic basis. Many nations do not like the idea of others knowing the nature and extent of their resource base better than they know it themselves. Efforts have been made to accommodate such nations. One suggested solution was turning off a resource satellite's sensors over certain areas. The small size and proximity of neighboring nations with divergent wishes is only one unrealistic aspect of this solution.

Nations which are unenthusiastic about others knowing their circumstances include many who nevertheless would themselves like to know their own resource base better. Efforts have been made by U.S. government agencies to offset the problem by having foreign nations participate in the activities. Training programs for foreign nationals are carried on and a number of the ERTS-1 research grants were awarded to people in other nations. Such actions have helped to keep the stature of the problem small, but they do not address the heart of the matter. The United Nations has a number of organizational units such as the Outer Space Affairs Division of the Secretariat and the Committee on the Peaceful Uses of Outer Space, and the latter has a legal subcommittee. If we reflect on the problems that have attended efforts to establish uniformity among coastal nations in claiming territorial jurisdiction of offshore waters, it shouldn't surprise us that a satisfactory manual of jurisprudence for outer space is not yet available.

A more personal aspect of the legal question comes to mind when we consider the implications of some of the capabilities that are forecast for remote sensing. If an individual person's image can, in fact, be resolved from sensors in an orbiting satellite, foreign *or* domestic, just how much detail can be expected? Will an individual be recognizable? It is difficult to imagine this possibility; for one thing, a top view is hardly ideal for identification. Somewhere along the line, however, a question of invasion of privacy could conceivably be raised, if not on an individual or intranational level, perhaps on the international level. Remote sensing offers a capability for minding other people's business on a scale never before approached. As of the moment, there is no assurance that we are developing legal checks and balances appropriate to this new capability.

What if. . .

A look into an imaginary future may be an effective way to bring home the sobering responsibility that accompanies the promise that remote sensing offers. For this purpose, let us imagine that all of the capabilities that seem indicated by experiments to date have become realities, and further, that these capabilities are extended by sophistication of equipment. (That some will not become reality is offset by others not presently known that will.) Imagine also that men will find the means to employ this tool on a scale that allows them to know the entire world in such detail that there are few secrets left undiscovered. Or to make such a thing a little more believable, let us restrict our consideration to the resource base. Over the past several decades discoveries of unknown mineral deposits have continued despite extensive investigation in preceding years. Although most of the obvious ones were discovered early and new ones require increasingly complicated search efforts, several very extensive ones have emerged relatively recently.

But what if we really knew how much of everything there

was and where it was? Suppose that we knew where 99 percent of this or that mineral resource was located, which has never before been the case. What would be the economic effect—or political—or social? What if our knowledge of the sea became so complete that the chance element was removed from fishing or the harvest of any of its resources? Or if we did know in advance what the world food crop would be? And so on. . .

In reality, of course, remote sensing will not do these things alone, and such a situation, if it will ever occur, is an indefinite distance in the future. And yet, remote sensing will advance our knowledge—though not equally or simultaneously on all fronts—faster than we have imagined possible. The implied question is whether or not man will be ready for such knowledge.

There always have been inadequately surveyed areas which, with a closer and better look, have yielded the additional reserves of nonrenewable resources to allow things to go on, more or less, as before. Or man's seemingly limitless ingenuity has provided substitutes which usually supply the need satisfactorily or better. And, after some pretty bad mistakes, we have had some success in stopping short of extermination of some species; a few have even experienced a modest recovery. Despite these considerations, and the fact that prophets have not by and large enjoyed a high rate of accuracy, there still are some who feel that we are in a different situation now. They point to the size and rate of increase of the world population, the likelihood of increased demand per capita, and the unprecedented current rate of resource consumption as the basis for their concern. As they consider the potential of such tools as remote sensing, they wonder if we are not approaching a closed system with finite limits instead of the open-ended one we have so long enjoyed. Long before these limits have been reached, the effect of approaching them will be felt—economically, politically, and socially. A further intriguing aspect of this line of thought is the question of who would make the decisions about how to handle such situations.

International Cooperation

Others see many of the above entries on the debit side of the remote sensing ledger differently. The immature state of the art may be fortuitous, for example, in view of the stage of development of international law governing its use. Similarly, the existence of spy satellites of various nationalities might well be a major deterrent to aggression. Indeed, it has been suggested that with the current advances in orbital surveillance capability, UN-directed inspection from space should offer a major contribution to arms limitation agreements (Davies and Murray, 1972). It is suggested also that these advances in fact leave no alternative to international cooperation and will thus promote it. Enough positive aspects of remote sensing have been presented in the previous chapters that we need not rebut fully each of the negative factors discussed above. However, some of the broader aspects of international cooperation deserve additional consideration.

The trend toward increasing utilization of space platforms for remote sensing will probably proceeed more smoothly as other nations become more involved. As satellites become more numerous and sensors more sophisticated, objections to them will increase unless the benefits for all peoples outweigh the undesirable aspects. But since the full potential of remote sensing cannot be realized without international involvement anyhow, the problem is limited in scope.

Despite the present efforts, substantial international involvement is a bigger task than is immediately apparent. ERTS-1 imagery is now available to foreign nationals, but is of little more than aesthetic value to those who do not know how to use it. The value of the tool must be demonstrated to the appropriate government heads effectively enough that they will want to set up the mechanisms for application. And then there is the task of setting up these mechanisms—which is at best only just beginning even in the United States.

There is a reciprocal relationship between the above problems and the ideal of international cooperation. Clearly, combining the human and material resources of many nations can help produce better and faster solutions. On the other hand, the need to work together on a task may actually lead to increased cooperation and mutual respect among participating nations. Space shuttles, for example, will probably both require and further such cooperation (Mueller, 1972). Smaller nations, desiring to utilize space technology more actively, might jointly produce second or third generation shuttles of their own using knowledge gained through their participation in first-stage experimental efforts. The more widespread use of shuttles by people other than astronauts should help spread the perception of Earth as the one and only habitat of all mankind.

Although we have made only limited reference to ecological

considerations in earlier chapters, it should be evident that this factor could become a catalyst for the promotion of global cooperation. Concern for environment never has been and is not now a uniquely American trait. In many parts of the world there is growing sentiment in favor of wiser, more careful use of the resource base. The realization also is becoming more widespread that many environmental problems cannot be dealt with effectively on a local or even national basis. Some forms of life range so widely over the globe that the effect of good management in one place is completely nullified by poor management in another. Also, oceanic and atmospheric circulation distribute some products of misuse far and wide. The evolution of remote sensing at this point in man's development is especially timely. Man now has the numbers and the requisite capability for profoundly altering the physical system in an undesirable fashion. Perhaps this tool will give us the information base we need to prevent inadvertent ecological disasters and the monitoring capability we need to prevent deliberate misuse of the environment.

Conclusion

Remote sensing is expensive and immature, it is the basis of conflict and is subject to misuse, and it promises problems of awesome responsibility; but it offers a way to know the world intimately, and that knowledge can do incalculable good. Its future use and development may see emphases substantially different from those which have been discussed here, but the dimensions have been portrayed correctly.

The value of remote sensing has now been demonstrated clearly enough that its further use is assured. Men typically do not turn away from exciting glimpses of the future. Having seen the possibilities this technique offers, we can never be satisfied with our current state of relative ignorance. Remote sensing will play a major part in an information revolution that could alter man's relationship with his environment as strikingly as our present relationship differs from that of stone age man.

Appendix

Imagery Sources

Space Imagery

The Geological Survey of the U.S. Department of Interior recently activated its new EROS Data Center near Sioux Falls, South Dakota. The facility is in operation and imagery now is obtainable. Although the center is scheduled to have far greater scope, it is notably the source of aircraft and spacecraft imagery from the ERTS program. It also will be a source of Skylab data.

You may order ERTS imagery in a variety of formats —microfilm, positive transparencies, prints—and at several scales. ERTS black and white transparencies (for each of the several bands the sensors received) or color composites are obtainable. At this writing, only selected area color images have been processed; for other areas some delay must be anticipated. If detail is desired, transparencies are typically superior to prints, although they require viewing equipment.

Because there are a number of factors in addition to those mentioned above that should be considered in ordering ERTS imagery, it may be advisable to write for instructions and forms before actually placing an order. This procedure will provide you with a current price list also. The address is:

EROS Data Center
Sioux Falls, SD 57198

Although Sioux Falls is the government ERTS data center, there also are commercial firms which obtain the basic data and offer it for sale in similar or additional formats. One such a firm is:

Earth Satellite Corporation
1747 Pennsylvania Ave., NW
Washington DC 20036

Quite a bit of photography was obtained during the several Apollo missions and some of the earlier Gemini missions. Government agencies made this photography available to commercial organizations and several offer the photography for sale in various formats. Color slides and color or black and white prints at different scales may be ordered from catalogues. One such a firm is:

Technology Applications Center
The University of New Mexico
Albuquerque, NM 87106

Apollo and Gemini imagery is scheduled to become available from the EROS Data Center also.

Although the imagery is of decidedly lower resolution, photography produced from weather satellite data also is obtainable. Weather satellites have provided data from a greater variety of sensors than ERTS or manned space flights (except for Skylab), however, and nonphotographic data (such as thermal IR) is included in this category. The quantity of data is enormous; catalogues for weather satellite data form a small library. To obtain weather satellite data, write to:

The National Climatic Center
National Oceanic and Atmospheric Administration
Federal Building
Asheville, NC 28801

Aircraft Imagery

Nonphotographic

This category of remote sensor imagery has consistently been the most difficult to obtain. Radar, passive microwave, thermal

IR, and ultraviolet data collection efforts have been notably oriented toward manufacturer's testing and sales promotion or to rather specific contact research programs. The likelihood that the type of data desired exists for a given area of interest thus is not very great. Still, much government-supported research has been carried on in recent years and it would be worth the serious student's time to investigate the availability of such data. The NASA aircraft program has provided a variety of types of data from overflights of areas covered by contract research effort. This data and the airphoto coverage, much of it color IR, will be obtainable from the EROS Data Center. An initial step would be to indicate the area of interest and data type desired and to request ordering instructions.

For the person who simply desires samples of imagery and for whom the timing and location of data collection are unimportant, one approach is writing to the manufacturers of the equipment. Most firms have a public relations office which is assigned the task of distributing examples of their product's capability. While the response varies among firms, most will respond positively to a request with merit. There is a large enough number of companies that to attempt to list them here would be unrealistic; they can be identified, however, from their advertisements in the trade and professional journals. A few such journals are included in the bibliography in this volume.

Photographic

As mentioned above, a considerable amount of aerial photography, including IR photography, has been obtained in connection with research projects supported by government grants; this photography will be obtainable through the EROS Data Center. Its coverage is spotty, however, as many parts of the United States were not the site of research projects.

Over the years, conventional photography of all parts of the country has been obtained at one time or another (for some areas repeatedly) for various government agencies. The Forest Service, the Soil Conservation Service, the Geological Survey, and the Agriculture Stabilization and Commodity Service are examples. The latter agency has the greatest coverage although there is coverage overlap among the agencies. Photography of some areas is obtainable back to the 1930s and there are variable scales obtainable in some of the recent photography. The bulk of it, however, is vertical black and white photography of fairly large (1:20,000) scale. In time some of this photography also is to become available from the EROS Data Center. There is such a variety of sources for this conventional photography that for the uninitiated purchaser, the best procedure, at present, is to write for ordering instructions from:

Map Information Office
U.S. Department of Interior
Geological Survey
Washington DC 20240

Bibliography

Chapter 1

Bylinsky, G. "From a High Flying Technology a Fresh View of Earth." *Fortune*, June 1968, pp. 100–103.

Colwell, R. N. "Remote Sensing of Natural Resources." *Scientific American*, January 1968, pp. 54–69.

Estes, J. E., and L. W. Senger, eds. *Remote Sensing: Techniques for Environmental Analysis*. Santa Barbara: Hamilton Publishing Company, 1973.

Holz, R. K., ed. *The Surveillant Science, Remote Sensing of the Environment*. Boston: Houghton Mifflin, 1973.

Weaver, K. F. "Remote Sensing: New Eyes to See the World." *National Geographic Magazine*, January 1969, pp. 46–73.

Chapter 2

Edgerton, E., and D. Trexler. "Oceanographic Applications of Remote Sensing with Passive Microwave Techniques." *Proceedings of the 6th Symposium on Remote Sensing of Environment*, pp. 767–773. Willow Run: University of Michigan, 1969.

Estes, J. E., and B. Golomb. "Oil Spills: Method for Measuring their Extent on the Sea Surface." *Science*, 14 August 1970, pp. 676–678.

Manual of Color Aerial Photography. Falls Church, Va.: The American Society of Photogrammetry, 1968.

Morain, S. A., and D. S. Simonett. "Vegetation Analysis with Radar Imagery." *Proceedings of the 4th Symposium on Remote Sensing of Environment*, pp. 605–622. Willow Run: University of Michigan, 1966.

Remote Sensing with Special Reference to Agriculture and Forestry. Washington, D.C.: National Academy of Sciences, 1970.

Simon, I. *Infrared Radiation*. New York: D. Van Nostrand Co., 1966.

Simonett, D. S. "Remote Sensing with Imaging Radar: A Review."
Geoforum 2 (1970): 61−73.

Chapter 3

Cowell, R. N. "Remote Sensing as an Aid to the Management of Earth
Resources." *American Scientist*, March−April 1973, pp. 175−183.

MacDonald, H. C. "Geologic Evaluation of Radar Imagery from Darien
Province, Panama." *Modern Geology* 1 (1969): 1−64.

Roberts, W. O. "We're Doing Something About the Weather." *National
Geographic Magazine*, April 1972, pp. 518−555.

Sabatini, R. R., G. A. Rabchevsky, and J. E. Sissala. *Nimbus Earth
Resources Observations* (Technical Report No. 2; Contract no. NAS 5-
21617). Greenbelt, Md.: NASA Goddard Space Flight Center, 1971.

Sabins, F. F., Jr. "Thermal IR Imagery and its Application to Structural
Mapping in Southern California." *Geological Society of America
Bulletin* (March 1969) 397-404.

Chapter 4

Alexander, R. H. "Central Atlantic Regional Ecological Test Site." *4th Annual Earth Resources Program Review*. Vol. 3, pp. 72-1 to 72-9.
Houston: NASA MSC, 1972.

Anderson, J. R., E. Hardy, and J. Roach. "A Land-Use Classification
System for Use with Remote Sensor Data". U. S. Geological Survey circular 671.

Horton, F. E., and D. F. Marble. "Housing Quality in Urban Areas: Data
Acquisition and Classification Through the Analysis of Remote Sensor
Imagery." *2nd Annual Earth Resources Program Review*, pp. 15-1 to
15-13. Houston: NASA MSC, 1969.

"Ground Observations and Utility Evaluations of Space and High-Altitude
Photography, Eastern Arizona"; prepared by Raytheon Company,
Autometric Operation; for U. S. Department of the Interior; July 1970.

MacPhail, D. D.; "Photomorphic Mapping in Chile"; *Photogrammetric
Engineering* 37 (November 1971), 1139−1148.

Moore, E. G. and B. S. Wellar; "Urban Data Collection by Airborne Sensor"; *American Institute of Planners Journal* 32 (January 1969), 35−43.

Park, A. B., R. N. Colwell and V. I. Myers; "Resource Survey by Satellite:
Science Fiction Coming True"; *Science For Better Living* (USDA Year-
book of Agriculture, 1968); pp. 13−19.

Peplies, R. W. and J. D. Wilson; "Analysis of a Space Photo of a Humid
Forested Region: A Case Study of the Tennessee Valley" (Association of
American Geographers, Remote Sensing Commission, Technical Report

70-6). E. Tennessee State University, 1970.

Remote Multispectral Sensing in Agriculture (Laboratory for Application of Remote Sensing Annual Report, vol. 4 [Research Bulletin #873]). Lafayette, Indiana: Purdue University, December 1970.

Remote Sensing with Special Reference to Agriculture and Forestry. Washington, D.C.: National Academy of Sciences, 1970.

Rudd, R. D.; "Macro Land-Use Mapping With Simulated Space Photos"; *Photogrammetric Engineering* 37 (April 1971), 365—372.

Thrower, N. J. W. and L. W. Senger; "Satellite Photography for Mapping Land Use of the Southwestern U. S." (Association of American Geographers, Remote Sensing Commission, Technical Report 69-3). E. Tennessee State University, 1970.

Chapter 5

Data Users Handbook, NASA ERT Satellite (Document No. 71SD4249). Prepared by Space Division, General Electric for Goddard Space Flight Center, NASA. Greenbelt, Md.: 1972.

Fink, D. E.; "Special Report: Skylab Mission, Experiments"; *Aviation Week and Space Technology*, 2 April 1973, pp. 38—44.

Skylab EREP Investigators Data Book. Houston: NASA MSC, 1972.

Strickland, Z. "Special Report: Earth Resources Technology Satellites"; *Aviation Week and Space Technology*, 31 July 1972, pp. 46—62.

Symposium on Significant Results Obtained From ERTS—1, Abstracts. Greenbelt, Md.: NASA Goddard Flight Center, 1973.

This Island Earth (NASA Special Publication #250). Washington, D. C.: Office of Technology Utilization, 1970.

Chapter 6

Congress, 92nd, 1st Session (1971). *For the Benefit of All Mankind: A Survey of the Practical Returns from Space Investment.* House Report 92—748.

Davies, M. and B. Murray; "Space Observations, Disarmament, and the UN" *Astronautics and Aeronautics* 10 (September 1972), 60—65.

Mueller, G.; "Space Shuttle: Beginning a New Era in Space Cooperation"; *Astronautics and Aeronautics* 10 (September 1972), 20—25.

Peaceful Uses of Earth-Observation Spacecraft, Volume II: Survey of Applications and Benefits, Willow Run: Infrared and Optical Sensor Laboratory (University of Michigan), February 1966.

Remote Sensing with Special Reference to Agriculture and Forestry. Washington, D.C.: National Academy of Sciences, 1970.

"Resource Satellite Effort Spurred"; *Aviation Week and Space Technology*, 17 November 1969, pp. 79—88.

Rosenberg, P.; "Resolution, Detectability, and Recognizability"; *Photogrammetric Engineering* 37 (December 1971), 1255—1258.

Index